D0354436

Heat

The Age of Consent
Captive State
No Man's Land
Amazon Watershed
Poisoned Arrows

GEORGE MONBIOT

Heat

How to Stop the Planet Burning

with research assistance from
DR MATTHEW PRESCOTT

ALLEN LANE
an imprint of
PENGUIN BOOKS

ALLEN LANE

Published by the Penguin Group
Penguin Books Ltd, 80 Strand, London WC2R 0RL, England
Penguin Group (USA) Inc., 375 Hudson Street, New York, New York 10014, USA
Penguin Group (Canada), 90 Eglinton Avenue East, Suite 700, Toronto, Ontario, Canada M4P 2Y3
(a division of Pearson Penguin Canada Inc.)
Penguin Ireland, 25 St Stephen's Green, Dublin 2, Ireland (a division of Penguin Books Ltd)
Penguin Group (Australia), 250 Camberwell Road,
Camberwell, Victoria 3124, Australia (a division of Pearson Australia Group Pty Ltd)
Penguin Books India Pvt Ltd, 11 Community Centre,
Panchsheel Park, New Delhi – 110 017, India
Penguin Group (NZ), cnr Airborne and Rosedale Roads, Albany,
Auckland 1310, New Zealand (a division of Pearson New Zealand Ltd)
Penguin Books (South Africa) (Pty) Ltd, 24 Sturdee Avenue,
Rosebank, Johannesburg 2196, South Africa

Penguin Books Ltd, Registered Offices: 80 Strand, London WC2R 0RL, England

www.penguin.com

First published 2006
1

Copyright © George Monbiot, 2006

The moral right of the author has been asserted

Set in 9.75/13 pt Linotype Sabon
Typeset by Rowland Phototypesetting Ltd, Bury St Edmunds, Suffolk
Printed in Great Britain by Clays Ltd, St Ives plc

A CIP catalogue record for this book is available from the British Library

hardback
ISBN-13: 978-0-713-99923-5
ISBN-10: 0-713-99923-3

trade paperback
ISBN-13: 978-0-713-99924-2
ISBN-10: 0-713-99924-1

To Hanna
May this be a fit world for you to inhabit

Contents

Introduction
The Failure of Good Intentions

The god thou servest is thine own appetite.
Doctor Faustus, Act II, Scene 1[1]

Two things prompted me to write this book. The first was something that happened in May 2005, in a lecture hall in London. I had given a talk about climate change, during which I had argued that there was little chance of preventing runaway global warming unless greenhouse gases were cut by 80 per cent.[2] The third question stumped me.

'When you get your 80 per cent cut, what will this country look like?'

I hadn't thought about it. Nor could I think of a good reason *why* I hadn't thought about it. But a few rows from the front sat one of the environmentalists I admire and fear most, a man called Mayer Hillman. I admire him because he says what he believes to be true and doesn't care about the consequences. I fear him because his life is a mirror in which the rest of us see our hypocrisy.

'That's such an easy question I'll ask Mayer to answer it.'

He stood up. He is 75, but looks about 50, perhaps because he goes everywhere by bicycle. He is small and thin and fit-looking, and he throws his chest out and holds his arms to his sides when he speaks, as if standing to attention. He was smiling. I could see he was going to say something outrageous.

'A very poor third-world country.'

At about the same time I was reading Ian McEwan's novel *Saturday*. Henry Perowne comes home from his game of squash and steps into the shower.

When this civilisation falls, when the Romans, whoever they are this time round, have finally left and the new dark ages begin, this will be one of the first luxuries to go. The old folk crouching by their peat fires will tell their disbelieving grandchildren of standing naked mid-winter under jet streams of hot clean water, of lozenges of scented soaps and of viscous amber and vermilion liquids they rubbed into their hair to make it glossy and more voluminous than it really was, and of thick white towels as big as togas, waiting on warming racks.[3]

Was I really campaigning for an end to all this? To ditch the comforts Perowne celebrates and which I – like all middle-class people in the rich world – now take for granted?

There are aspects of this civilization I regret. I hate the lies and the political corruption, the inequality, the export of injustice, the military adventures, the destruction of wild places, the noise, the waste. But in the rich nations most people, most of the time, live as all prior generations have dreamt of living. Most of us have a choice of work. We have time for leisure, and endless diversions with which to fill it. We may vote for any number of indistinguishable men in suits. We may think and say what we want, and though we might not be heeded, nor are we jailed for it. We may travel where we will. We may indulge ourselves 'up to the very limits imposed by hygiene and economics'. We are, if we choose to be, well-nourished. Women – some women at any rate – have been released from domestic servitude. We expect effective healthcare. Our children are educated. We are warm, secure, replete, at peace.

For the first two million years of the history of the genus *Homo*, we lived according to circumstance. Our lives were ruled by the vicissitudes of ecology. We existed, as all animals do, in fear of hunger, predation, weather and disease.

For the following few thousand years, after we had developed a rudimentary idea of agriculture and crop storage, we enjoyed greater food security, and soon destroyed most of our non-human predators. But our lives were ruled by the sword and the spear. We fought, above all, for land. We needed it not just to grow our crops but also to provide power – grazing for our horses and bullocks, wood for our fires.

Then we began to discover some of the opportunities afforded by fossil fuels. No longer were we constrained by the need to live on ambient energy; we could support ourselves by means of the sunlight stored – in the form of carbon – over the preceding 350 million years. The new fuels permitted the economy to grow – to grow sufficiently to absorb some of the people dispossessed by the previous era's land disputes. Industry and cities boomed. Forced together within the workplace and the warren, the dispossessed could start to organize. The despots empowered by the seizure of land were forced to loosen their grip.

Fossil fuels helped us to fight wars of a horror never contemplated before, but they also reduced the need for war. For the first time in human history – indeed for the first time in biological history – there was a surplus of available energy. We could survive without having to fight someone for the resources we needed. Our freedoms, our comforts, our prosperity are all the products of fossil carbon, whose combustion creates the gas carbon dioxide, which is primarily responsible for global warming. Ours are the most fortunate generations that have ever lived. Ours might also be the most fortunate generations that ever will. We inhabit the brief historical interlude between ecological constraint and ecological catastrophe.

Oh, those distant, sunny days of May 2005, when I believed this problem could be solved with a mere 80 per cent cut! After my talk, a man called Colin Forrest wrote to me. I had failed, he explained, to take note of the latest projections. He sent me a paper he had written whose argument (which I will explain at greater length in the next chapter) I could not fault.[4]

If in the year 2030, carbon dioxide concentrations in the atmosphere remain as high as they are today, the likely result is two degrees centigrade of warming (above pre-industrial levels). Two degrees is the point beyond which certain major ecosystems begin collapsing. Having, until then, absorbed carbon dioxide, they begin to release it. Beyond this point, in other words, climate change is out of our hands: it will accelerate without our help. The only means, Forrest argues, by which we can ensure that there is a high chance that the temperature does not rise to this point is for the rich nations to cut their greenhouse gas emissions by 90 per cent by 2030. This is the task whose feasibility *Heat* attempts to demonstrate.

By 'feasibility' I mean compatibility with industrial civilization. Within the environmental movement there are some people who regard the preservation of this state as an unworthy goal. The slogan of North American EarthFirst!, for example, is 'Back to the Pleistocene'. But even if you would prefer to be running around in skins, chasing or being chased by giant aurochs, advocating a return to the economy of the Stone Age is futile, for the great majority of people find this prospect unappealing. Even demanding the restitution of a largely agricultural society, or the economy of 'a very poor third-world country' would be pure self-indulgence. Whether or not we enjoy the soft life (and I suspect that some of those who advocate its dissolution would be among the first to perish in the wilderness), it is politically necessary to discover the means of sustaining it. This book seeks to devise the least painful means of achieving a 90 per cent cut in carbon emissions. It attempts to reconcile our demand for comfort, prosperity and peace with the restraint required to prevent us from destroying the comfort, prosperity and peace of other people. And though I began the search for these solutions almost certain that I would be unsuccessful, I now believe it can be done.

Heat is both a manifesto for action and a thought experiment. Its experimental subject is a medium-sized industrial nation: the United Kingdom. It seeks to show how a modern economy can be de-carbonized while remaining a modern economy. Though the proposals in this book will need to be adjusted in countries with different climates and of greater size, I believe the model is generally applicable: if the necessary cut can be made here, it can be made by similar means almost anywhere.

I concentrate on the rich nations for this reason: until we have demonstrated that we are serious about cutting our own emissions, we are in no position to preach restraint to the poorer countries. The rich world's most common excuse for inaction can be expressed in one word: China. It is true that China's emissions per person have been rising by around 2 per cent a year.[5] But they are still small by comparison to our own. A citizen of China produces, on average,

2.7 tonnes of carbon dioxide a year. A citizen of the United Kingdom emits 9.5, and of the United States, 20.0.[6] To blame the Chinese for the problem, and to claim that their rapacious appetites render our efforts futile, is not just hypocritical. It is, I believe, another manifestation of our ancient hysteria about the Yellow Peril.

After looking at what the impacts of unrestrained climate change might be, and at why we have been so slow to respond to the threat, I begin my search for solutions within my own home. I show how years of terrible building, feeble regulations and political cowardice have left us with houses scarcely able to perform their principal function, which is keeping the weather out. I look at the means by which our existing homes could be redeemed and better ones could be built, and discover what the physical and economic limits of energy efficiency might be.

I then seek to determine how best their energy might be supplied. Before I began my research on that subject, I thought it would be quite easy to cover: I would need only decide whether we should use wind, waves or solar power, or nuclear energy, or biomass, or a means of stripping carbon dioxide from the exhausts of power stations. But the more I read, the more difficult and contradictory the questions became. The three chapters dealing with this issue are the most technically complex in the book. I believe – though by the skin of my teeth – that I might have found a workable solution.

Next I show how a new system for land transport could cut carbon emissions by 90 per cent with scarcely any reduction in our mobility. But when I come to examine aviation, I discover that there are simply no effective technological solutions: in this chapter I have failed in my attempt to reconcile the luxuries we enjoy with the survival of the biosphere, and I am forced to conclude that the only possible answer is a massive reduction in flights.

Then I look at two industrial sectors – retailing and cement manufacture, both of which produce disproportionate amounts of carbon dioxide – and propose some radical means by which shops can stay in business and houses can be built without melting the ice caps. I have tried throughout this account to identify the methods that are cheapest, that have already been shown to work and that are most compatible with the lives we lead already.

I would like to believe that the changes I suggest could be achieved by appealing to people to restrain themselves. But though some environmentalists, undismayed by the failure of the past forty years of campaigning, refuse to see it, self-enforced abstinence alone is a waste of time.

What is the point of cycling into town when the rest of the world is thundering past in monster trucks? By refusing to own a car, I have simply given up my road space to someone who drives a hungrier model than I would have bought. Why pay for double-glazing when the supermarkets are heating the pavement with the hot air blowers above their doors? Why bother installing an energy-efficient lightbulb when a man in Lanarkshire boasts of attaching 1.2 million Christmas lights to his house? (Mr Danny Meikle told journalists that he needs two industrial meters to measure the electricity he uses. One year his display melted the power cable supplying his village.[7] The name of the village – which proves, I think, that there is a God – is Coalburn.)

And which of us – except perhaps Mayer Hillman – can really claim to live as we urge others to live? Most environmentalists – and I include myself in this – are hypocrites. I know of a British climate-change campaigner who spends her holidays snorkelling in the Pacific, and she doesn't get there by bicycle. One friend – a prominent environmentalist – burns coal on an open fire. Another – a biodiversity campaigner – serves tuna steaks to his guests. In an interview with the *Guardian* conducted in Las Vegas, Chris Martin, the lead singer of Coldplay, spoke about the songs on his album *X&Y*.

Twisted Logic is an intense, angry track encouraging people to make the right decisions about how they live their lives and how they treat the planet.[8]

A few paragraphs later, he revealed that he was about to

fly by private jet to Palm Springs, 35 minutes from Las Vegas. The band can now afford to fly wherever possible, and the increased privacy and speed mean that Apple will be able to join her father on tour more often. 'I certainly don't want her to stay at home all the time,' Martin says. 'As she gets older, hopefully she'll come out as and when she wants. I always thought it'd be cool to be in school and say, "I'm not coming in today –

I'm off to Costa Rica to see my dad play." I do think that wins you a few points.'[9]

At the beginning of his *Organic Bible*, the green gardener Bob Flowerdew explains that organic gardening means 'minimizing ecological damage and making best use of resources'.[10] He goes on to boast that 'when most people are only planting their [new potatoes] on Good Friday, as is traditional in the UK, I am eating mine.'[11] How? By growing them in a heated greenhouse.

We might buy eco-friendly washing-up liquid and washable nappies. But we cancel out any carbon savings we might have made ten thousand-fold whenever we step on to an aeroplane. Our efforts are tokenistic. By and large, whatever our beliefs might be, we consume as much as our incomes allow. Environmentalism is for other people.

What this means is that changes of the kind I advocate in this book cannot take place without constraints which apply to everyone, rather than to everyone else. I am sorry to say that only regulation – that deeply unfashionable idea – can quell the destruction wrought by the god we serve, the god of our own appetites. Manmade global warming cannot be restrained unless we persuade the government to force us to change the way we live.

I have mentioned that one of the gifts fossil fuels have granted us is freedom: freedom to choose how we should live, to go where we wish, to buy what we want. A 90 per cent cut in our emissions of carbon dioxide is, I admit, an inherently narrow constraint. I did not invent it – it is what the science appears to demand. But within that constraint, we should be free to live as we wish. The need to tackle climate change must not become an excuse for central planning. The role of government must be to establish the limits of action, but to guarantee the maximum of freedom within those limits. And it must help us by ensuring that even within those constraints, life remains as easy as possible. In Chapter 3 I explain how this might best be done.

I am not writing this book to confirm what you believe to be true. Many of the things I say will disturb and upset people who have taken an interest in this subject. As always, I seem destined to offend everyone. But I am sorry to report that an extraordinary amount of rubbish has been written by well-meaning people about tackling climate change. It is hard to see how it helps us to pretend that certain measures work when they do not.

Let me give you an example. In 2005 the environmental architect Bill Dunster, who designed the famous BedZed zero-carbon development outside London, published a brochure purporting to show how homes could best be refurbished. 'Up to half of your annual electric needs,' it claimed, 'can be met by a near silent micro wind turbine.'[12] The turbine he specified has a diameter of 1.75 metres.[13] He suggested it be attached to the gable end of the house. It looks like a bargain, as it costs only £1000.

Later that year the magazine *Building for a Future*, which supports renewable energy, published an analysis of micro wind turbines. It found that a 1.75 metre turbine would produce about 5 per cent of a household's annual electricity demand.[14] To provide the 50 per cent Bill Dunster advertises, you would need a turbine 4 metres in diameter.[15] If you attached a beast like this to the gable end of your house, the lateral thrust it exerted would rip the building to bits. Though it did not say as much, the magazine's analysis made it clear that micro wind turbines are a waste of time and money. In most environmental circles this admission is heresy.

One of the discoveries I have made in writing this book is that my instincts are almost always wrong. Like many environmentalists I have succumbed, for example, to what could be described as the aesthetic fallacy: I have made the mistake of confusing what is aesthetically pleasing with what is environmentally sound. For instance, I have always assumed that candles are more environmentally friendly than electric lighting, for no better reason than that I like them and that they produce less light. In his excellent textbook on energy systems, Godfrey Boyle points out that in terms of the light given off per watt of expended power, a candle is 71 times less efficient than an old-fashioned incandescent bulb, and 357 times worse than a compact fluorescent model.[16] The same applies to oil lamps. Boyle notes that

It is quite remarkable that the complex process of choosing to burn a litre of kerosene in an engine, to drive a generator, to power a fluorescent lamp, can produce 250–450 times more useful light than burning the same amount in an oil lamp.[17]

Nothing here is as it seems. The research for this book has involved me in a long series of surprises. I am sure that they will continue long after it is published, as my findings and proposals are challenged and refined by others. But what I have sought to do throughout the text is to start from first principles, to believe nothing until it is demonstrated, to junk any technology, however pleasing it may be, which does not work. What I am attempting to do is to find the least painful means of making real cuts, rather than the least painful means of being seen to do something.

One of the hardest tasks I have faced is deciding whom to trust. Many of those who have written about climate change have economic interests in the outcome. In some cases, as I will show in Chapter 2 (The Denial Industry), these interests have been heavily disguised: the oil companies, for example, speak with many voices. On the other side, environmentalists – as the example I have given suggests – have often made wild claims unsupported by verifiable facts. In some cases such claims support their own economic interests, though these are generally undisguised. One rule I have devised for myself is to trust no one who has something to sell. By tracing the statements different people have made back to their roots, I have developed a kind of heirarchy of credibility.

When trying to decide which solutions work and which ones don't, the organizations I have found most useful are learned societies and special committees – such as the Royal Commission on Environmental Pollution, the House of Lords Science and Technology Committee and the House of Commons Environmental Audit Committee – and academic institutions, such as Oxford University's Environmental Change Institute, the Tyndall Centre on Climate Change, the UK Energy Research Centre and the US National Academy of Engineering. Their reports draw together hundreds of years of collective experience. The International Energy Agency and the US Energy Information Administration, though partisan, are useful sources of raw data. Rather to my surprise, given that it has become so closely associated with spin and the massaging of figures, I have also found most of the British government's technical reports to be reliable: the data seem to be manipulated only *after* they have been collected. For news about technological developments, I've found *New Scientist*, *Energy World* and *Building for a Future* especially helpful.

When attempting to determine what climate change will do to the planet, the choice, at first sight, seems simpler: the most credible sources are peer-reviewed academic journals, and particularly the most illustrious ones, such as *Science* and *Nature*. But the science – as science always should be – is contradictory and confusing. There is no 'answer'; simply a story with many tellers, which changes every day. From time to time, committees of scientists try to reach an overview. The most eminent of these, bringing together thousands of researchers, is the Intergovernmental Panel on Climate Change (IPCC), which produces an 'assessment report' every few years. Another useful summary was provided by a conference run by the UK's Meteorological Office in 2005, which tried to work out the total impacts of climate change on different ecosystems and human populations.

But not all the topics I have investigated have been covered by these distinguished bodies. In some important respects they have abandoned us. It has been left to amateurs to try to perform the carbon-cutting calculation I explain in Chapter 1, and to work out a fair method of deciding how the right to pollute should be allocated. None of the official reports I have read will tell you how much electricity a micro wind turbine produces or, for that matter, what percentage of our electricity can be generated by wind or wave or solar power without causing the national grid to collapse. So I have been forced either to rely on less august sources or to try to work out the answers for myself.

In other cases there is too much data, by which I mean that the bodies I have learnt to trust have produced conflicting estimates, and I have no means of deciding which one should be believed. This is especially true when it comes to the costs of energy, over which there is a remarkable degree of dispute. In these cases, I have published a range of estimates.

*

I have one purpose in writing this book: to persuade you that climate change is worth fighting. I hope I have been able to demonstrate that it is not – as some people (notably the geophysiologist James Lovelock) have claimed – too late. In doing so, I hope to prompt you not to lament our governments' failures to introduce the measures required to tackle it, but to force them to reverse their policies, by joining what must become the world's most powerful political movement.

Failing all that, I have one last hope: that I might make people so depressed about the state of the planet that they stay in bed all day, thereby reducing their consumption of fossil fuels.

1

A Faustian Pact

The framing of this circle on the ground
Brings whirlwinds, tempests, thunder and lightning
Doctor Faustus, Act II, Scene 1[1]

There was more than one Faust. The name, which means 'the fortunate' in Latin, was used by German magicians much as conjurors today might call themselves 'the magnificent' or 'the incredible'. But we know which one he was. In 1513 in Erfurt a Conrad Mudt heard an 'immoderate and Foolish Braggart' describe himself as the 'demigod from Heidelberg'.[2] His name was 'Georg Faust'. In 1528, a 'Jörg Faust' was thrown out of the town of Ingolstadt, and in 1532 a 'Dr Faust, the great sodomite and necromancer' was denied entry to Nuremberg.[3] People were plainly afraid of him. When he died in Württemberg in 1540 or 1541, the locals claimed that the Devil had taken him home.

After his death, his story began to spread, and in 1587 an amplified version was published by an anonymous theologian in Frankfurt.[4] Two years later it was translated into English as *The History of the Damnable Life and Deserved Death of Doctor John Faustus*. This was the source for Christopher Marlowe's play *The Tragical History of Doctor Faustus*, which appears to have been written in 1590.

Marlowe tells the story of a brilliant scholar, 'glutted ... with learning's golden gifts',[5] who reaches the limits of human knowledge. Bored by terrestrial scholarship, he plots, by means of necromancy, to break into

1

> . . . a world of profit and delight
> Of power, honor, of omnipotence.[6]

When, he believes, he has acquired his demonic powers, spirits will fetch him everything he wants:

> I'll have them fly to India for gold,
> Ransack the ocean for orient pearl,
> And search all corners of the new-found world
> For pleasant fruits and princely delicates.[7]

So Faustus draws a circle and summons the Devil's servant, Mephistopheles. He offers him a deal: if the Devil will grant him twenty-four years in which to 'live in all voluptuousness',[8] Faustus will, at the end of that period, surrender his soul to hell. Mephistopheles explains the consequences, but the doctor refuses to believe him.

> Think'st thou that Faustus is so fond to imagine
> That, after this life, there is any pain?
> Tush, these are trifles and mere old wives' tales.[9]

So the bargain is struck and signed in blood, and Faustus acquires his magical powers. With the help of a flying 'chariot burning bright', he takes a sightseeing tour around Europe. He performs miracles. He summons fresh grapes from the southern hemisphere in the dead of winter. After twenty-four years, the devils come for him. He begs for mercy, but it is too late. They drag him down to hell.

If you did not know any better, you could mistake this story for a metaphor of climate change.

Faust is humankind, restless, curious, unsated. Mephistopheles, who appears in the original English text as 'a fiery man',[10] is fossil fuel. Faust's miraculous abilities are the activities fossil fuel permits. Twenty-four years is the period – about half the true span – in which they have enabled us to live in all voluptuousness. And the flames of hell – well, I think you've probably worked that out for yourself.

In 1590 the economy was powered largely by wood, water, wind and horses. The English did burn some fossil fuel: we know, for example, that in 1585 London imported about 24,000 tons of coal.[11]

That coal would have provided as much energy as the United Kingdom now consumes in half an hour.*

Liquid fossil fuels were not to be widely used for almost three centuries. Europe was submerged in the Little Ice Age: temperatures were one to one and a half degrees cooler than they are today. Science, with a few exceptions, was a muddle of alchemy, theology and magic. If man-made climate change had taken place by then, the people of the sixteenth century would have had no means of detecting it. *The Tragical History of Dr Faustus* is not an allegory of climate change. But the intention of the poet does not affect the power of the metaphor. Our use of fossil fuels is a Faustian pact.

To doubt, today, that manmade climate change is happening, you must abandon science and revert to some other means of understanding the world: alchemy perhaps, or magic.

Ice cores extracted from the Antarctic show that the levels of carbon dioxide and methane in the atmosphere (these are the two principal greenhouse gases) are now higher than they have been for 650,000 years.[14,15] Throughout that period, the concentration of these gases has been closely tracked by global temperatures.[16]

Carbon dioxide (CO_2) levels have been rising over the past century faster than at any time over the past 20,000 years.[17] The only means by which greenhouse gases could have accumulated so swiftly is human action: carbon dioxide is produced by burning oil, coal and gas and by clearing forests, while methane is released from farms and coal mines and landfill sites.[18]

Both gases let in heat from the sun more readily than they let it out. As their levels in the atmosphere increase, the temperature rises. The concentration of carbon dioxide, the more important of the two, has risen from 280 parts per million parts of air (ppm) in Marlowe's time to 380 ppm today.[19] Most of the growth has taken place in the

---

* The government's Department of Trade and Industry gave the UK's use of coal in 2003 as 68.7 million short tons.[12] One short ton = 2000lb; one ton = 2240lb. 68.7 ÷ 2240 x 2000 = 61.3m tons. The US Energy Information Administration reports that coal provides a 15 per cent share of the UK's total energy consumption[13] 61.3 ÷ 15 x 100 = 408.7mt. 408,700,000 ÷ 24,000 = 17,029. In one year, there are 8,760 hours. 8,760 ÷ 17,029 = 0.514.

past fifty years. The average global temperature over the past century has climbed, as a result, by 0.6° centigrade.* According to the World Meteorological Organization, 'the increase in temperature in the twentieth century is likely to have been the largest in any century during the past 1000 years'.[20]

If you reject this explanation for planetary warming, you should ask yourself the following questions:

1. Does the atmosphere contain carbon dioxide?

2. Does atmospheric carbon dioxide raise the average global temperature?

3. Will this influence be enhanced by the addition of more carbon dioxide?

4. Have human activities led to a net emission of carbon dioxide?

If you are able to answer 'no' to any one of them, you should put yourself forward for a Nobel Prize. You will have turned science on its head.

But the link has also been established directly. A study of ocean warming over the past forty years, for example, published in the journal *Science* in 2005, records a precise match between the distribution of heat and the intensity of manmade carbon dioxide emissions.[21] Its lead author described his findings thus:

The evidence is so strong that it should put an end to any debate about whether humanity is causing global warming.[22]

This sounds like a strong statement, but he is not alone. In 2004, another article in *Science* reported the results of a survey of scientific papers containing the words 'global climate change'.[23] The author found 928 of them on the database she searched. 'None of the papers,' she discovered,

disagreed with the consensus position ... Politicians, economists, journalists and others may have the impression of confusion, disagreement, or discord among climate scientists, but that impression is incorrect.[24]

---

* All temperatures in this book are expressed in centigrade.

In 2001 the Royal Society, the United Kingdom's pre-eminent scientific institution, published the following statement:

Despite increasing consensus on the science underpinning predictions of global climate change, doubts have been expressed recently about the need to mitigate the risks posed by global climate change. We do not consider such doubts justified.[25]

It was also signed by the equivalent organisations in fifteen other countries.*

Similar statements have been published by the US National Academy of Sciences,[26] the American Meteorological Society,[27] the American Geophysical Union[28] and the American Association for the Advancement of Science.[29]

Until 2005, there was one remaining line of evidence permitting some people to claim that manmade climate change could still be disputed. A study of satellite measurements conducted in 1992 by the atmospheric scientists Roy Spencer and John Christy found that part of the atmosphere (the lower troposphere) had cooled over the preceding thirteen years.[30] This, in a warming world, should not have been possible. In 2005, three separate studies showed that the data had been misread.[31,32,33] Professor Christy admitted that his results were incorrect and agreed that the atmosphere had warmed. As the author of one of the studies pointed out,

there is no longer any data contradicting the predictions of global warming models.[34]

Already sea ice in the Arctic has shrunk to the smallest area ever recorded.[35] In the Antarctic, scientists watched stupefied in 2002 as the Larsen B ice shelf collapsed into the sea.[36] A paper published in

---

* The Australian Academy of Sciences, the Royal Flemish Academy of Belgium for Sciences and the Arts, the Brazilian Academy of Sciences, the Royal Society of Canada, the Caribbean Academy of Sciences, the Chinese Academy of Sciences, the French Academy of Sciences, the German Academy of Natural Scientists Leopoldina, the Indian National Science Academy, the Indonesian Academy of Sciences, the Royal Irish Academy, the Accademia Nazionale dei Lincei (Italy), the Academy of Sciences Malaysia, the Academy Council of the Royal Society of New Zealand, the Royal Swedish Academy of Sciences.

Science concluded that its disintegration was the result of melting caused by a warming ocean.[37] The global sea level has been rising by around 2 millimetres a year,[38] partly because water expands as it warms, partly because of the melting of ice and snow.

Almost all the world's glaciers are now retreating.[39,40] Permafrost in Alaska and Siberia, which has remained frozen since the last Ice Age, has started to melt.[41,42] Parts of the Amazon rainforest are turning to savannah as the temperatures there exceed the point at which trees can survive.[43] Coral reefs in the Indian Ocean and the South Pacific have begun to wilt. The World Health Organization estimates that 150,000 people a year are now dying as a result of climate change, as diseases spread faster at higher temperatures.[44] All this is happening with just 0.6° of warming.

The Intergovernmental Panel on Climate Change (IPCC), a committee of climate specialists which assesses and summarizes the science, estimated in 2001 that global temperatures will rise by between 1.4 and 5.8° this century.[45] Since then, some climate scientists have come to believe that this range is too low: one study published in 2005, for example, suggests that the maximum possible temperature rise which could be caused by a doubling of carbon dioxide concentrations is 11.5°.[46] An increase as big as this, however, is very unlikely.

But even a much smaller rise is likely to cause great harm to some human populations. Professor Martin Parry of the UK's Meteorological Office estimates that a rise of just 2.1° will expose between 2.3 and 3 billion people to the risk of water shortages.[47] The disappearance of glaciers in the Andes and the Himalayas will imperil the people who depend on their meltwater, particularly in Pakistan, western China, Central Asia, Peru, Ecuador and Bolivia.[48,49] As rainfall decreases, there are likely to be longer and more frequent droughts in southern Africa, Australia and the countries surrounding the Mediterranean.[50] In northern Europe, summer droughts and winter floods will both become more frequent. Very wet winters, for example, which until now have troubled us every forty years or so, could recur one year in every eight.[51]

The UN Food and Agriculture Organisation warns that

in some forty poor, developing countries, with a combined population of two billion . . . [crop] production losses due to climate change may drastically increase the number of undernourished people, severely hindering progress in combating poverty and food insecurity.[52]

The reason is that, in many parts of the tropics, crop plants are already close to their physiological limits. If, for example, temperatures stay above 35° for one hour while rice is flowering, the heat will sterilize the pollen.[53] The International Rice Research Institute has found that rice yields fall by 15 per cent with every degree of warming.[54]

When I first read about this, I thought it equated to a formula for worldwide famine, and said as much in the *Guardian*. I was wrong to do so. Climate scientists, I later discovered, were confident that lower crop yields in some parts of the tropics would be offset by higher crop yields in temperate countries.[55] In the cooler parts of the world, the productive season lengthens and both higher temperatures and higher carbon dioxide levels should allow crop plants to grow faster.

But now, I am sorry to say, it seems that I might have been right, though for the wrong reasons. In late 2005, a study published in the *Philosophical Transactions of the Royal Society* alleged that the yield predictions for temperate countries were 'over optimistic'.[56] The authors had blown carbon dioxide and ozone, in concentrations roughly equivalent to those expected later this century, over crops in the open air. They discovered that the plants didn't respond as they were supposed to: the extra carbon dioxide did not fertilize them as much as the researchers predicted, and the ozone reduced their yields by 20 per cent.[57] Ozone levels are rising in the rich nations by between 1 and 2 per cent a year, as a result of sunlight interacting with pollution from cars, planes and power stations. The levels happen to be highest in the places where crop yields were expected to rise: western Europe, the midwest and eastern US and eastern China. The expected ozone increase in China will cause maize, rice and soybean production to fall by over 30 per cent by 2020. These reductions in yield, if real, are enough to cancel out the effects of both higher temperatures and higher carbon dioxide concentrations.[58,59]

Another paper in the same journal pointed out that, as carbon dioxide levels rise, plants release less water from their leaves.[60] This reduces local rainfall, which in many regions will have declined already because of climate change. The result, which has not been anticipated in the standard climate models, could be a further decline in crop production. It now seems possible that the world could be pushed towards famine.

The effects of crop losses are likely to be compounded by other problems. Though this prediction is controversial, some scientists suggest that, as temperatures rise, the incidence of malaria will increase. One study maintains that temperatures 2.3° higher than today's will expose a further 180–230 million people to the risk of catching the disease.[61] Diarrhoea and cholera are both associated with rising temperatures.[62,63]

If the earth warms by a moderate amount and sea levels increase by some 40 centimetres (roughly in the middle of the expected range for this century), the number of people in danger of saltwater floods caused by storm surges could grow from some 75 million (today) to around 200 million.[64] As the sea rises, salt water will pollute the drinking water on which some of the biggest coastal cities – Shanghai, Manila, Jakarta, Bangkok, Kolkata, Mumbai, Karachi, Lagos, Buenos Aires and Lima – depend.[65] In some cases, according to the International Association of Hydrogeologists, this problem could be big enough to necessitate the cities' abandonment.[66]

The West Antarctic Ice Sheet contains enough water to raise sea levels by a further 3 metres,[67] enough to inundate parts of New York, London, Tokyo, Mumbai, indeed of most of the world's major cities. The ice sheet appears to be starting to disintegrate.[68] There is great controversy about how long this process will take. The sheet is propped up by ice shelves extending into the sea, like a roof kept aloft by the walls of a house. If they collapse as the Larsen B did, the ice sheet could begin to slide into the ocean. No one knows how swiftly this would happen, but it is unlikely that the entire sheet could dissolve in less than 300 years. If just 10 per cent of it fell into the sea this century, the results would be catastrophic for many coastal peoples.

When the IPCC produced its last overview of the science of climate change, it found that 'there was no compelling evidence' to indicate that storms in and around the tropics had become worse.[69] But in 2005 two papers, published in *Science* and *Nature*, suggested that the intensity of hurricanes had increased since the mid 1970s.[70,71] It is not yet clear whether this is connected to climate change, though there is a relationship between the temperature of the sea surface and the strength of a storm.[72] In March 2004 the first hurricane ever recorded in the South Atlantic hit the coast of Brazil.

The number of extreme weather events of all kinds appears to have quintupled since the 1950s, according to the insurance company Munich Re.[73] The summer of 2003 seems to have been the hottest in Europe since at least the year 1500.[74] Thousands of people in Europe and India died as a result of the heatwave. According to a paper published in *Nature*, human influence has at least doubled the chances of its recurrence.[75] In northern Europe, however, the number of people dying because of extreme temperatures is likely to drop, as our winters become warmer.[76]

Other species will be hit sooner and harder than humans. In 2004 researchers on five continents surveyed the ecosystems covering 20 per cent of the earth's surface. They found that, if temperatures rise to about the middle of the expected range, between 15 and 37 per cent of the world's species are 'committed to extinction' by 2050.[77] With just 1.4° of warming, the coral reefs in the Indian Ocean will become extinct.[78] With 2°, some 97 per cent of the world's reefs will bleach – which means the coral animals eject the algae which keep them alive, and are likely to die as a result.[79] As increasing levels of carbon dioxide dissolve in seawater, the oceans will acidify. Their pH could fall from 8.2 to 7.7 by the end of the century,[80] and by 2050 the water could become too acid for shells to form. This will be devastating to sea life, wiping out much of the plankton upon which the marine ecosystem depends. With 2° of warming, all the sea ice in the Arctic could melt in summer, killing the polar bears, the walruses and much of the rest of the ecosystem.[81]

In one of the most depressing papers I have ever read, researchers from University College London and the Met Office reported in 2005 that 'the Amazonian forest is currently near its critical resiliency threshold.' With just a small degree of warming 'the interior of the Amazon Basin becomes essentially void of vegetation.'[82]

The problem is that the trees in some parts of the forest are responsible for as much as 74 per cent of local rainfall.[83] As they start to die when the temperature rises, less water is released into the air by the forest. This has three effects: there is less rainfall to sustain the remaining trees, more sunlight reaches the forest floor (drying it and making the forest more susceptible to fires), and less heat is lost through evaporation. The rising temperature and decreasing rainfall kill more trees, and the chain reaction continues. It could happen soon and swiftly: 'we suggest,' the researchers say, 'that this threshold exists very near to current climatic conditions.'[84]

The Amazon is the most biodiverse place on earth, but the problem does not stop with other species. It produces the rain which sustains much of South America. And trees, roughly speaking, are sticks of wet carbon. As they burn or rot – as they oxidize in other words – they turn into carbon dioxide. The Amazon has the potential to release 730 million tonnes of carbon – about 10 per cent of manmade emissions – a year for seventy-five years.[85]

This is just one of the means by which climate change begets climate change. A paper published in *Geophysical Research Letters* in 2003 predicted that, as a result of global warming, by about 2040 living systems on the land will start to release more carbon dioxide than they absorb. By 2100, it suggests, the surface of the earth will be emitting around 7 billion tonnes of carbon a year,[86] which is roughly what human beings produce today. This is an example of 'positive feedback': climate change accelerating itself. Positive feedback was not fully considered by the IPCC when it predicted that the temperature would rise by between 1.4 and 5.8°.[87]

One of the reasons why the terrestrial biosphere begins to release more carbon dioxide than it absorbs is that, as we have seen, plants in the tropics and even some temperate regions[88] may shrivel or die when the temperature rises. But there are several others. Soil, for example, becomes a net source of carbon when temperatures rise, as the metabolism of the microbes it contains speeds up. This was not supposed to happen for several decades,[89] but in 2005 British scientists reported that soils in England and Wales had already become carbon sources.[90] The carbon dioxide they were releasing had cancelled out all the cuts that the UK had made since 1990. Before the end of the century, the world's soils will eject the manmade carbon they have absorbed over the past 150 years.[91]

As the permafrost in the far north melts, it starts to release methane. The West Siberian bog alone, which began melting in 2005, is believed to contain 70 billion tonnes of the gas,[92] whose liberation would equate to 73 years of current manmade carbon dioxide emissions.*

The National Center for Atmospheric Research in the US estimates that 90 per cent of the top 10 feet of permafrost throughout the Arctic could thaw by 2100.[95] These positive feedbacks – and there are many more – extend the possible range of global temperatures. In doing so, they make a truly catastrophic event more likely to happen.

One such event has seized the imagination of people in northern Europe. The region is kept warm in the winter – relative to parts of the world at the same latitude – by the northwards transport of water from the Caribbean – a current known as the Gulf Stream. The Gulf Stream is part of a general oceanic circulation, which is mostly driven by the sinking of surface waters in the far north of the Atlantic. As they roll southwards over the seabed, they create the currents which, after a long journey, return to northern Europe, carrying heat from the tropics.

The reason they sink is that they are both cold and salty, and therefore denser than the waters beneath them. The phenomenon is known as 'thermohaline circulation', or THC. For at least twenty years, some oceanographers have warned that this sinking, and therefore the 'overturning circulation' (the deep ocean currents which drive the whole system), could either weaken or stop altogether because meltwater flowing into the Arctic seas would dilute the salty surface waters. If this happened, northern Europe could be reduced to tundra, while the tropics, as heat was not transported away from them, would become very much hotter. This has taken place before. As the northern hemisphere began to warm after the last Ice Age, the ice dam holding back a vast lake in North America burst. The freshwater thundering into the north Atlantic appears to have shut down

---

* Methane has a warming effect 23 times as great as carbon.[93] Manmade carbon dioxide emissions are currently around 22 billion tonnes a year (this is 3.667 × the weight of the carbon they contain).[94]

the ocean circulation, with the result that temperatures in Europe fell by 5°. They did not recover for 1,300 years.

Many climate scientists believe that a total shutdown of this nature is impossible: there is simply not enough freshwater in the far north to prevent the surface waters from sinking.[96] At most, a slightly weakened current might reduce the rate of warming in northern Europe. In July 2005, the British House of Lords examined the evidence for the possible impacts of climate change and concluded that 'changes in the THC are not at all likely to occur, as we understand it, in the next 100 years.'[97] This was a reasonable summary of the existing science.

Five months later, *Nature* reported

the first observational evidence that ... a decrease of the oceanic overturning circulation is well underway.[98]

Researchers from the National Oceanography Centre in the United Kingdom claimed to have discovered that the circulation had in fact been weakening for fifty years, but that it had not hitherto been detected.[99] It appears to have slowed down by 30 per cent.

At the same time, the overflow waters and in turn the deep waters of the North Atlantic have significantly freshened ... Increased freshwater input into the Nordic Seas will initially weaken the circulation only slowly. But when a certain threshold is reached, the circulation may jump abruptly to a new state in which there is little or no heat flux to the north.[100]

If this occurs, it would have

devastating effects on socio-economic conditions in the countries bordering the eastern North Atlantic.[101]

The possible switch from one stable state (a smoothly flowing Gulf Stream) to another (no Gulf Stream at all) is an example of what climate scientists call 'non-linearity'. They point out that some of the earth's systems are unlikely to respond smoothly to changes in the climate: they could flip suddenly from one condition to another.[102]

I have concentrated so far on the effects which could take place within the IPCC's range of 1.4–5.8° of global warming. But there are, as I have mentioned, some climate scientists who maintain that the temperature this century could rise much further.

The Nobel laureate Paul Crutzen, having taken into account the falling levels of particles produced by heavy industry in the atmosphere, which have so far sheltered us from some of the sun's heat, has made a rough estimate that the temperature could rise by between 7 and 10°.[103] In 2005, British scientists published the results of a computer simulation larger and more detailed than its predecessors. It revealed that a doubling of carbon dioxide concentration in the atmosphere could lead to temperatures ranging anywhere from 1.9 and 11.5° above their pre-industrial levels.[104] This does not mean that all temperatures in this range are equally likely – the extremes are much less probable than the temperatures in the middle – but the researchers found that none of them could be ruled out.[105]

So what happens if average global temperatures rise by more than 6°? There could be a historical precedent.

The Permian period came abruptly to an end 251 million years ago. In China, South Africa, Australia, Greenland, Russia and Spitsbergen, the rocks record the same sequence of events, taking place almost instantaneously.[106] The marine sediments deposited at the time show two sudden changes. The red or green rock laid down in the presence of oxygen is replaced by black muds of the kind deposited when oxygen is absent. An instant shift in the ratio of the isotopes (alternative forms) of carbon within the rocks suggests a very rapid change in the concentration of atmospheric gases. On land, gently deposited mudstones and limestones give way to great dumps of pebbles and boulders.

The Permian was one of the most biologically diverse periods. Sabre-toothed reptiles hunted herbivores the size of rhinos through forests of tree ferns and flowering trees. Among the coral reefs lived great sharks, fish of all kinds and hundreds of species of shelly creatures. At the point at which the sediments change, 251 million years ago, the fossil record very nearly stops dead. The reefs die instantly, and do not reappear on earth for 10 million years. All the large and medium-sized sharks disappear, most of the shelly species, even the majority of the plankton. Among many classes of marine animals, the only survivors were those adapted to the near-absence of oxygen.[107]

Plant life was almost eliminated from the earth's surface. The four-footed animals, the group to which humans belong, were nearly exterminated: so far only two fossil reptile species have been found anywhere on earth which survived the end of the period. The world's surface came to be dominated by just one of these, which was about the size and shape of a pig. It became ubiquitous because nothing else was left to compete with it or to prey upon it. Altogether, some 90 per cent of the earth's species appear to have been wiped out:[108,109] this represents by the far the biggest of the mass extinctions. The world's 'productivity' (the total mass of biological matter) collapsed.

These events coincided with a series of volcanic eruptions in Siberia; the eruptions which gave rise to the Siberian Traps. The volcanoes produced great quantities of two gases: sulphur dioxide and carbon dioxide. These gases appear to have caused the extinctions. The sulphur and other effusions caused acid rain, but would have bled from the atmosphere quite quickly. The carbon dioxide, on the other hand, persisted. The rising temperatures caused by the gas appear to have warmed the world sufficiently to have destabilized a super-concentrated form of methane which was found then (and is still found today) in large quantities in the sediments beneath the polar seas. The release of methane into the atmosphere might explain the sudden shift in carbon isotopes. The temperature rose by between 6°[110] and 8°.[111]

Ocean circulation appears to have dropped, for reasons which will now be familiar to you, to about one twentieth of current levels,[112] depriving the deeper waters of oxygen. As the plants on land died, their roots would no longer have held the soil and loose rock together, with the result that erosion rates greatly increased.

This does not mean that we can make a direct comparison between the events which brought the Permian to an end and the possible effects of manmade climate change today. Many of the plants on land were doubtless killed by acid rain rather than by high temperatures. Though some countries seem to be doing their best to replicate both conditions, sulphur emissions are much lower today than they were 251 million years ago. But it does give us an indication of the possible scale of ecological change a temperature rise of this magnitude could provoke.

Various other outcomes of climate change have been proposed, of which the most intriguing is one suggested by a reader of mine.

Thank you for drawing attention to the threat of global warming. I wish the world would wake up to how serious it is. If we don't do something soon the whole planet could turn into a dessert.

This is a tempting prospect, but I regret to say that the science does not support it.

Curtailing climate change must, in other words, become the project we put before all others. If we fail in this task, we fail in everything else. But is it possible? Is it, as James Lovelock sometimes suggests,[113] too late?

I don't believe it is. We have a short period – a very short period – in which to prevent the planet from starting to shake us off. Our aim must be to stop global average temperatures from rising to more than 2° above pre-industrial levels, which means more than 1.4° above the current point.

Two degrees, because it has been widely recognized by climate scientists as the critical threshold,[114,115] has sometimes been characterized as a 'safe' level of warming. As I hope this account has shown, it is merely less dangerous than what lies beyond. A conference of scientists convened by the UK's Met Office warned that at less than 1° above pre-industrial levels, crop yields begin to decline in continental interiors,[116] droughts spread in the Sahel region of Africa,[117] water quality falls and coral reefs start to die.[118] At 1.5° or less, an extra 400 million people are exposed to water stress and another 5 million to hunger,[119] 18 per cent of the world's species will be lost[120] and the 'onset of complete melting of Greenland ice' is triggered.[121] There are, I am afraid, some effects of climate change which cannot be avoided.

Two degrees is important because it is the point at which some of the larger human impacts and the critical positive feedbacks are expected to begin. If we do not greatly reduce our emissions, temperatures are likely to reach that point in about 2030.[122]

My correspondent Colin Forrest, who is not a professional climate scientist but appears to have done his homework, argues his case as follows. Researchers at the Potsdam Institute for Climate Impact in Germany have estimated that holding global temperatures to below 2° means stabilizing concentrations of greenhouse gases in the atmosphere at or below the *equivalent* of 440 parts of carbon dioxide per million.[123] While the carbon dioxide concentration currently stands at 380 parts, the other greenhouse gases raise this to an equivalent of 440 or 450. In other words, if everything else were equal, greenhouse gas concentrations in 2030 would need to be roughly the same as they are today.

Unfortunately, everything else is not equal. By 2030, according to a paper published by scientists at the Met Office, the total capacity of the biosphere to absorb carbon will have reduced from the current 4 billion tonnes a year to 2.7 billion.[124] To maintain equilibrium at that point, in other words, the world's population can emit no more than 2.7 billion tonnes of carbon a year in 2030. As we currently produce around 7 billion, this implies a global reduction of 60 per cent. In 2030, the world's people are likely to number around 8.2 billion. By dividing the total carbon sink (2.7 billion tonnes) by the number of people, we find that to achieve stabilization the weight of carbon emissions per person should be no greater than 0.33 tonnes. If this problem is to be handled fairly, everyone should have the same entitlement to release carbon, at a rate no greater than 0.33 tonnes per year.

In the rich countries, this means an average cut by 2030 of around 90 per cent. The United Kingdom, for example, currently releases 2.6 tonnes per capita,*[125] so would need to reduce its emissions by 87 per cent. Germany requires a cut of 88 per cent, France of 83 per cent, the United States, Canada and Australia 94 per cent.†[126] By contrast, the Kyoto Protocol to the United Nations Framework Convention on Climate Change – the only international agreement that has been struck so far – commits its signatories to cut their carbon emissions by a total of 5.2 per cent by 2012.

---

* This measures just the carbon in carbon dioxide. To obtain the weight of CO_2, you must multiply this figure by 3.667, which gives the UK's emissions a value of 9.5 tonnes.
† This assumes that the other greenhouse gases, such as methane, nitrous oxide, hydrofluorocarbons and sulphur hexafluoride, are cut at the same rate.

These could be underestimates. The Potsdam Institute calculates that with the equivalent of 440 ppm of carbon dioxide in the atmosphere, there is a 67 per cent chance of holding the temperature rise to below 2°.[127] Another study suggests that to obtain a 90 per cent chance of stabilization below 2°, you would need to keep the concentration below 400 parts per million – 40 or 50 parts below the current level.[128] Because the carbon released now stays in the atmosphere for some 200 years.[129] and causes climate change many years into the future, there is perhaps a 30 per cent chance that we have already blown it. We might already be committed to 2°.

But I am writing this book in the spirit of optimism, so I refuse to believe it.

Whether or not it is too late to hold global temperatures below the critical threshold, it is clear that the greater the cuts we make, the lesser the eventual impact will be. A 90 per cent cut should make the sort of warming that took place at the end of the Permian impossible. It is also clear that the sooner we act, the more effective the cut will be. There are several reasons for this, but the most obvious is illustrated by the two graphs on p. 18. In both cases we reach the target of a 90 per cent reduction by 2030, but in the second graph, where we delay the cut for longer, our total emissions are higher.

Two centuries after the *Tragical History of Doctor Faustus* was published, Johann Wolfgang von Goethe rewrote the magician's story. In his version – *Faust* – the doctor's bargain with Mephistopheles changes. He offers Mephistopheles his soul, but on one further condition: hell can have him only if he stops striving and succumbs to 'smug complacency'.[130]

> You heard me, there can be no thought of joy.
> Frenzy I choose, most agonizing lust,
> Enamored enmity, restorative disgust.[131]

Faust acquires his powers and performs his miracles, but he never relaxes. As the story progresses, he becomes less interested in living in all voluptuousness and begins pouring his demonic energies into other schemes. Towards the end of his life he starts planning a development project. He will create 'room to live for millions', sheltered

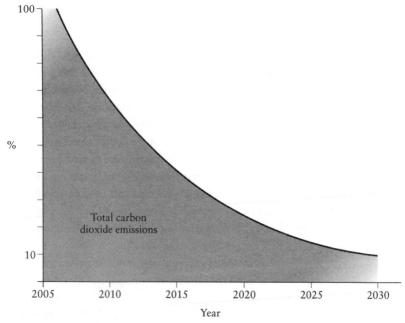

1. Fast carbon reduction

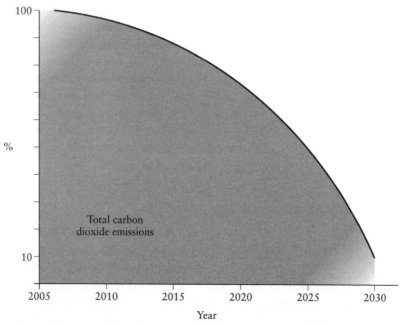

2. Slow carbon reduction

Carbon emissions as a percentage of the current total

from the storms and tides.[132] He will use wave power to provide energy for them and human ingenuity to rescue land from the sea. He dies in the midst of his labours, and Mephistopheles is cheated of his prey. Angels descend and bear Faust's soul up to heaven.

Faust, in other words, is redeemed by working, with frenzy and agonizing lust (and, I am sorry to say, a good deal of brutality), for the greater human good. While he still possesses his dark powers – his command of technology and labour, his ability to effect political and economic change – he uses them to create a world in which a free and comfortable society can persist. The gifts which threatened to destroy him are deployed instead to save him. This book seeks to explain how it might best be done.

2

The Denial Industry

Look at the canting holy-oilers!
Thus they have snatched from us so many a prize,
With our own weapons they would foil us;
They too are devils, only in disguise.

Faust, Part II, Act V[1]

But first I want to examine why we have been so slow to act.

On almost every other serious issue, the professional classes appear to be better informed than the rest of the population. On climate change the reverse seems to be true. The only people I have met over the past three years who haven't the faintest idea what manmade climate change is or how it is caused are university graduates. In 2004, for example, I had to tell a press officer at the British government's Department for Transport what carbon dioxide was. In 2005, I heard an insurer explain that he was failing to persuade the financial markets to take climate change seriously, as the senior managers either hadn't understood it or didn't believe in it.[2] In 2006, I spoke to a journalist of 20 years' standing about the problem of rising carbon emissions, and was asked 'what are carbon emissions and why are they a problem?' But over the same period – and perhaps I don't get out enough – I have never spoken to a shop assistant, taxi driver, bar tender or vagrant who did not possess at least a vague idea of what climate change means and why it is happening.

From this I deduce that the problem is not that people aren't hearing about it, but that they don't want to know. The professional classes have the most freedom to lose and the least to gain from an attempt to restrain it. The effort to tackle climate change suffers from the problem of split incentives: those who are least responsible for it are the most likely to suffer its effects.

Bangladesh and Ethiopia are two of the countries which will be hit hardest. A sea level rise of 1 metre could permanently flood 21 per cent of Bangladesh, including its best agricultural land, pushing some 15 million people out of their homes.[3] Storm surges of the kind the country experienced in 1998 are likely to become more common: in that instance, 65 per cent of Bangladesh was temporarily drowned, and its farming and infrastructure ruined.[4] By 2004, half of Bhola, the country's largest island, on which 1.6 million people live, had already been washed away. Climate scientists blamed this on rising sea levels: the erosion rate has accelerated since the 1960s.[5]

Ethiopia has already been suffering a series of droughts linked to climate change. A paper published in the *Philosophical Transactions of the Royal Society*[6] shows that the spring rains have steadily diminished since 1996. It blames the trend on rising sea surface temperatures in the Indian Ocean.[7] In 2005, partly as a result of the droughts caused by the failure of these rains, between 8 and 10 million Ethiopians were at risk of starvation.

Most of the rich countries, being located in temperate latitudes, will, in the initial stages at least, suffer lesser ecological effects. They will also have more money with which to protect their citizens from floods, droughts and extremes of temperature. Within these countries, the richest people, who can buy their way out of trouble, will be harmed last. The blame, as this table suggests, is inversely proportional to the impacts.[8]

country	carbon dioxide emissions, 2003 (tonnes per capita)
Luxembourg	24.3
United States	20.0
United Kingdom	9.5
Bangladesh	0.24
Ethiopia	0.06

Source: US Energy Information Administration.[9]

The Ethiopians, on average, emit one 400th of the carbon dioxide produced by the people of Luxembourg, the country which has the highest gross domestic product per person.[10]

So asking wealthy people in the rich nations to act to prevent climate change means asking them to give up many of the things they value – their high-performance cars, their flights to Tuscany and Thailand and Florida – for the benefit of other people.

The problem is compounded by the fact that the connection between cause and effect seems so improbable. By turning on the lights, filling the kettle, taking the children to school, driving to the shops, we are condemning other people to death. We never chose to do this. We do not see ourselves as killers. We perform these acts without passion or intent.

Many of those things we have understood to be good – even morally necessary – must also now be seen as bad. Perhaps the most intractable cause of global warming is 'love miles': the distance you must to travel to visit friends and partners and relatives on the other side of the planet. The world could be destroyed by love.

To make this even more difficult, the early effects of climate change, for those of us who live in the temperate countries of the rich world, are generally pleasant. Our winters are milder, our springs come sooner. We have suffered the occasional flood and drought and heat-wave. But the overwhelming sensation, just when we need to act with the greatest urgency, is that of being blessed by our pollution.

A wealthy society's split incentives are shared by its government. As Tony Blair has remarked, 'there is a mismatch in timing between the environmental and electoral impact.'[11] By the time the decisions he has made come home to roost, he will have been out of office for years. If a government allows the growth of air travel to continue, for example, the effects are delayed, diffuse and hard to blame on any one source. If, by contrast, it restricts or reverses the growth in flights, the effects are immediately attributable to its actions. Everyone knows who is responsible if we may no longer fly to Thailand.

But it is not just a matter of a failure to engage. Assisting our reluctance is an active campaign of dissuasion.

I first became aware of it after reading a series of truly idiotic articles in the British press. In some newspapers, as the following examples suggest, a total absence of scientific knowledge is no barrier to publication.

George W. Bush is right. The Kyoto Treaty is a silly waste of time. The greenhouse effect probably doesn't exist. There is as yet no evidence for it (Peter Hitchens, *Mail on Sunday*).[12]

And if the climate is indeed overheating, that does not mean that manmade emissions are necessarily to blame. Indeed, it is extremely unlikely that they would be since carbon dioxide forms a relatively small proportion of the atmosphere, most of which consists of water vapour (Melanie Phillips, *Daily Mail*).[13]

The greenhouse effect, first observed in the middle of the nineteenth century, is the phenomenon which keeps this planet warm enough to sustain life. Peter Hitchens appears to have confused it with manmade climate change. But this did not prevent him from feeling qualified to continue thus:

Global warming is probably caused by, would you believe it, the sun. The temperature of the atmosphere, measured by NASA, has not risen in the past 22 years. There was global warming between 1870 and 1940, when there were far fewer greenhouse gases than now. The only reason these facts are so little-known is that a self-righteous love of 'the environment' has now replaced religion as the new orthodoxy . . . Why are we so blinkered to these lies?[14]

If most of the atmosphere consisted of water vapour, we would need gills. But Melanie Phillips is sure enough of her atmospheric physics to allege that

. . . the theory that global warming is all the fault of mankind is a massive scam based on flawed computer modelling, bad science and an anti-Western ideology . . . The majority of well-meaning opinion in the Western world believes a pack of lies and propaganda.[15]

At first I mistook all this for native idiocy, and doubtless this plays a supporting role. But it was after I investigated another set of assertions that I began to see that these claims did not originate in the newspapers.

Unlike most of those who have maintained in the media that climate change is not happening, David Bellamy is, or was, a scientist, formerly a senior lecturer in botany at the University of Durham. He was also an environmentalist and a famous and rather wonderful television presenter. In the early 2000s, he decided that climate change wasn't happening. Here is what he wrote in an article published in 2004 in the *Daily Mail*, under the title 'Global Warming? What a Load of Poppycock!'

The link between the burning of fossil fuels and global warming is a myth. It is time the world's leaders, their scientific advisers and many environmental pressure groups woke up to the fact.[16]

In April 2005, I read a letter of his in *New Scientist*.

Further to your coverage of climate change and melting ice in the Himalayas, it should be pointed out that glaciers in many other parts of the world are not shrinking but in fact are growing . . . Indeed, if you take all the evidence that is rarely mentioned by the Kyotoists into consideration, 555 of all the 625 glaciers under observation by the World Glacier Monitoring Service in Zurich, Switzerland, have been growing since 1980.[17]

I was astonished by this claim: it conflicted with everything I had read about 'glacier mass balance' – the degree to which they are advancing or retreating. So I telephoned the World Glacier Monitoring Service and read out Bellamy's letter. 'This,' they told me, 'is complete bullshit.'[18] The latest studies show unequivocally that most of the world's glaciers are retreating.[19]

But Bellamy's figures must have come from somewhere, so I e-mailed him to ask for his source. After several requests, he explained that he had found them on a website called www.iceagenow.com. I urge you to visit it: it is an extraordinary production. But there indeed was all the material Bellamy had cited in his letter, including the figures – or something resembling the figures – he quoted.

Since 1980, there has been an advance of more than 55 per cent of the 625 mountain glaciers under observation by the World Glacier Monitoring group in Zurich.'*[21]

The source, which Bellamy also cited in his e-mail to me, was given as 'the latest issue of 21st Century Science and Technology'.

This, I found, is a publication belonging to the American millionaire Lyndon Larouche. Larouche has claimed that the British royal family is running an international drugs syndicate,[22,23,24] that Henry Kissinger is a communist agent,[25] that the British government is controlled by Jewish bankers,[26] and that modern science is a conspiracy against human potential.[27] In 1989 he received a fifteen-year sentence for conspiracy, mail fraud and tax-code violations.[28,29]

21st Century Science and Technology gave no source for its figures; but the same data could be found all over the internet. They were first published online by the 'Science and Environmental Policy Project', which is run by an environmental scientist called Dr S. Fred Singer. After they were posted on his website – www.sepp.org – they were reproduced by several other groups, such as the Competitive Enterprise Institute,[30] the National Center for Public Policy Research[31] and The Advancement of Sound Science Coalition.[32] They had also found their way into the Washington Post.[33] But where did they come from? Fred Singer cited half a source:

a paper published in Science in 1989.[34]

I went through every edition of Science published in 1989, both manually and electronically. Not only did it contain nothing resembling those figures; throughout that year there was no paper published in this journal about glacial advance or retreat. Satisfied that the figures were nonsense, I left it there.

---

* You may have noticed that while Bellamy's source claimed that 55 per cent of 625 glaciers are advancing, Bellamy claimed that 555 of them – or 89 per cent – are advancing. This figure seems to have existed nowhere else. But on the standard English keyboard, '5' and '%' occupy the same key. If you try to hit '%', but fail to press shift, you get 555, instead of 55%. This is the only explanation I can produce for his figure. When I challenged him, he admitted that there had been 'a glitch of the electronics'.[20]

But after I had published these findings in the *Guardian*,[35] one of my readers wrote to Dr S. Fred Singer.

Dear Professor Singer,

How do you answer the statement by George Monbiot, writing in the *Guardian* newspaper on Tuesday, that you cited a non-existent paper in an unspecified 1989 edition of *Science* as the only source for a claim that most of the world's glaciers were advancing?[36]

Singer's response was interesting, and unexpected.

Monbiot is confused – or simply lying ... he has a vivid imagination that borders on being slanderous ... I know nothing about a 1989 paper in *Science*. This is the same Monbiot who along with other 'climate campaigners' complained in *Nature* (31 March 2005) that some of us are creating the impression that 'climate scientists are deeply divided' while – acc. to him – there is a 'robust' consensus about anthropogenic global warming. Obviously, he has been smoking something or other.[37]

My correspondent wrote back:

Dear Professor Singer,

Thank you for your quick reply to my query. However, I found your statements very puzzling ... When I did a Google search of www.sepp.org, I did find two pages that made exactly the same claim that Monbiot ascribed to you. It seems that he was neither lying nor confused ... Could you please be specific about this 1989 paper in *Science*? It looks very unlikely that the WGMS would have said what you claim.[38]

This time Singer replied in less aggressive tones: the claim, he said, had been posted on his site by 'former SEPP associate Candace Crandall'. It 'appears to be incorrect and has been updated'.[39] He forgot to add that Candace Crandall was his wife. Almost a year later, when writing this book, I checked his website, and found this paragraph:

The World Glacier Monitoring Service in Zurich, in a paper published in *Science* in 1989, noted that between 1926 and 1960, more than 70 per cent of 625 mountain glaciers in the United States, Soviet Union, Iceland, Switzerland, Austria and Italy were retreating. After 1980, however, 55 per cent of these same glaciers were advancing.[40]

It had not been changed. What I also found, on the SEPP site and the others which had published the glacier figures, were most of the allegations, however daft or misleading, which had later been made in the press by David Bellamy, Peter Hitchens, Melanie Phillips, the novelist Michael Crichton and most of the other prominent repudiators of manmade climate change. The groups I've listed appear to have compiled and distributed the facts and figures the writers used. The groups have something else in common: they have all been funded by Exxon.*[19]

ExxonMobil is the world's most profitable corporation. In autumn 2005, it reported *quarterly* profits of almost $10 billion, the highest corporate earnings on record.[41] It makes most of this money from oil, and has more to lose than any other company from efforts to tackle climate change. Its approach to the issue could be summed up thus:

Should the public come to believe that the scientific issues are settled, their views about global warming will change accordingly. Therefore, you need to continue to make the lack of scientific certainty a primary issue in the debate.

These words are not mine. Nor are they Exxon's. They were written for Republican Party activists by a political consultant named Frank Luntz, during the first mid-term election campaign in George W. Bush's presidency.[42] As we will see, there are plenty of places in which Exxon finds itself at home. But in other respects it has difficulties: it must confront a scientific consensus as strong as that which maintains that smoking causes lung cancer or that HIV causes AIDS.

The website www.exxonsecrets.org, using data found in the company's official documents, lists 124 organizations which have taken money from the company or work closely with those which have. They take a consistent line on climate change: that the science is contradictory, the scientists are split, environmentalists are charlatans,

---

* This is not to suggest that Bellamy, Hitchens, Phillips or Crichton have themselves taken any money from Exxon.

liars or lunatics, and if governments took action to prevent global warming, they would be endangering the global economy for no good reason. The findings these organizations dislike are labelled 'junk science'. The findings they welcome are labelled 'sound science'.

Among the organizations that have been funded by Exxon are some well-known websites and lobby groups, such as TechCentralStation, the Cato Institute and the Heritage Foundation. Some of those on the list have names which make them look like grassroots citizens' organizations or academic bodies: the Center for the Study of Carbon Dioxide and Global Change, for example; the National Wetlands Coalition; the National Environmental Policy Institute; the American Council on Science and Health.[43] One or two of them, such as the Congress of Racial Equality and George Mason University's Law and Economics Center, *are* citizens' organizations or academic bodies, but the line they take on climate change is very much like that of the other sponsored groups. While all these groups are based in the United States of America, their publications are read and cited, and their staff are interviewed and quoted, all over the world.

By funding a large number of organizations, Exxon helps to create the impression that doubt about climate change is widespread. For those who do not understand that scientific findings cannot be trusted if they have not appeared in peer-reviewed journals, the names of these institutes help to suggest that serious researchers are challenging the consensus.

This is not to claim that all the science these groups champion is bogus. On the whole, they use selection, not invention. They will find one contradictory study – such as the discovery of tropospheric cooling I mentioned in the previous chapter and which, in a garbled form, was used by Peter Hitchens – and promote it relentlessly. They will continue to do so long after it has been disproved by further work. Though, for example, John Christy, the author of the troposphere paper, admitted in August 2005 that his figures were incorrect,[44] his initial findings are still being circulated and championed by many of these groups, as a quick internet search will show you.

But they do not stop there. The chairman of Fred Singer's Science and Environmental Policy Project (Singer is the president) is a man called Frederick Seitz. Seitz is a physicist, who in the 1960s was president of the US National Academy of Sciences. In 1998, he wrote a document, known as the 'Oregon Petition', which has been cited by almost every journalist who claims that climate change is a myth.

The petition reads as follows:

We urge the United States government to reject the global warming agreement that was written in Kyoto, Japan, in December, 1997, and any other similar proposals. The proposed limits on greenhouse gases would harm the environment, hinder the advance of science and technology, and damage the health and welfare of mankind.

There is no convincing scientific evidence that human release of carbon dioxide, methane, or other greenhouse gasses is causing or will, in the foreseeable future, cause catastrophic heating of the Earth's atmosphere and disruption of the Earth's climate. Moreover, there is substantial scientific evidence that increases in atmospheric carbon dioxide produce many beneficial effects upon the natural plant and animal environments of the Earth.[45]

Anyone with a degree could sign it. It was attached to a letter written by Seitz, entitled *Research Review of Global Warming Evidence*:

Below is an eight-page review of information on the subject of 'global warming', and a petition in the form of a reply card. Please consider these materials carefully.

The United States is very close to adopting an international agreement that would ration the use of energy and of technologies that depend upon coal, oil, and natural gas and some other organic compounds.

This treaty is, in our opinion, based upon flawed ideas. Research data on climate change do not show that human use of hydrocarbons is harmful. To the contrary, there is good evidence that increased atmospheric carbon dioxide is environmentally helpful.

. . . Frederick Seitz. Past President, National Academy of Sciences.[46]

The lead author of the 'review' which followed Frederick Seitz's letter is a Christian fundamentalist called Arthur B. Robinson. He has never worked as a climate scientist.[47] It was co-published by Robinson's organization – the Oregon Institute of Science and Medicine – and an outfit called the George C. Marshall Institute, which has received $630,000 from ExxonMobil since 1998.[48] The other three authors

were Arthur Robinson's 22-year old son[49] and two employees of
the George C. Marshall Institute.[50,51] The chairman of the George
C. Marshall Institute was Frederick Seitz.[52]

The paper maintained that

As coal, oil, and natural gas are used to feed and lift from poverty vast
numbers of people across the globe, more carbon dioxide will be released
into the atmosphere. This will help to maintain and improve the health,
longevity, prosperity, and productivity of all people. . . . We are living in
an increasingly lush environment of plants and animals as a result of the
carbon dioxide increase. Our children will enjoy an Earth with far more
plant and animal life than that with which we now are blessed. This is a
wonderful and unexpected gift from the Industrial Revolution.[53]

It was printed in the font and format of the *Proceedings of the
National Academy of Sciences*: the journal of the organization of
which Frederick Seitz – as he had just reminded his correspondents –
was once president.

Soon after the petition was published, the National Academy of
Sciences released this statement:

The Council of the National Academy of Sciences is concerned about the
confusion caused by a petition being circulated via a letter from a former
president of this Academy . . . The petition was mailed with an op-ed article
from *The Wall Street Journal* and a manuscript in a format that is nearly
identical to that of scientific articles published in the *Proceedings of the
National Academy of Sciences*. The NAS Council would like to make it clear
that this petition has nothing to do with the National Academy of Sciences
and that the manuscript was not published in the *Proceedings of the National
Academy of Sciences* or in any other peer-reviewed journal. The petition does
not reflect the conclusions of expert reports of the Academy.[54]

But it was too late. Seitz, the Oregon Institute and the George
C. Marshall Institute had already circulated tens of thousands of
copies, and the petition had established a major presence on the
internet. Some 17,000 graduates signed it, the great majority of
whom have no background in climate science. It has been repeatedly
cited – by David Bellamy, Melanie Phillips and the rest – as a petition
by climate scientists. It is promoted by the Exxon-sponsored sites

as evidence that there is no scientific consensus on climate change.

All this is now well-known to climate scientists and environmentalists. But the most interesting thing I have discovered while researching this issue is that the corporate campaign to deny that manmade climate change is taking place was not initiated by Exxon, or by any other firm directly involved in the fossil fuel industry. It was started by the tobacco company Philip Morris.

In December 1992, the US Environmental Protection Agency published a 500-page report called *Respiratory Health Effects of Passive Smoking*. It found that

... the widespread exposure to environmental tobacco smoke (ETS) in the United States presents a serious and substantial public health impact.

In adults: ETS is a human lung carcinogen, responsible for approximately 3,000 lung cancer deaths annually in US nonsmokers.

In children: ETS exposure is causally associated with an increased risk of lower respiratory tract infections (LRIs) such as bronchitis and pneumonia. This report estimates that 150,000 to 300,000 cases annually in infants and young children up to 18 months of age are attributable to ETS.[55]

Had it not been for the settlement of a major class action against the tobacco companies in the United States, we would never have been able to see what happened next. But in 1998 they were forced to publish their internal documents and post them on the internet.

Within two months, Philip Morris, the world's biggest tobacco firm, had devised a strategy for dealing with the passive-smoking report. In February 1993 Ellen Merlo, its senior vice president of corporate affairs, sent a letter to William I. Campbell, Philip Morris's chief executive officer and president, explaining her intentions.

Our overriding objective is to discredit the EPA report ... Concurrently, it is our objective to prevent states and cities, as well as businesses, from passive-smoking bans.[56]

To this end, she had hired a public relations company called APCO. She had attached the advice it had given her. APCO warned that

No matter how strong the arguments, industry spokespeople are, in and of themselves, not always credible or appropriate messengers.[57]

So the fight against a ban on passive smoking had to be associated with other people and other issues. Philip Morris, APCO said, needed to create the impression of a 'grassroots' movement – one that had been formed spontaneously by concerned citizens to fight 'over-regulation'.[58] It should portray the danger of tobacco smoke as just one 'unfounded fear' among others, such as concerns about pesticides and cellphones.[59] APCO proposed to set up

a national coalition intended to educate the media, public officials and the public about the dangers of 'junk science'. Coalition will address credibility of government's scientific studies, risk assessment techniques and misuse of tax dollars . . . Upon formation of Coalition, key leaders will begin media outreach, e.g., editorial board tours, opinion articles, and brief elected officials in selected states.[60]

APCO would found the coalition, write its mission statements, and 'prepare and place opinion articles in key markets.' For this it required $150,000 for its own fees and $75,000 for the coalition's costs.[61]

By May 1993, as another memo from APCO to Philip Morris shows, the fake citizens' group had a name: The Advancement for Sound Science Coalition, or TASSC.[62] It was important, further letters stated, 'to ensure that TASSC has a diverse group of contributers';[63] to 'link the tobacco issue with other more "politically correct" products'; and to associate scientific studies which cast smoking in a bad light with 'broader questions about government research and regulations', such as

- Global warming
- Nuclear waste disposal
- Biotechnology.[64]

APCO would engage in the

intensive recruitment of high-profile representatives from business and industry, scientists, public officials, and other individuals interested in promoting the use of sound science.[65]

By September 1993, APCO had produced a 'Plan for the Public Launching of TASSC'.[66,67] The media launch would not take place in

Washington, DC or the top media markets of the country. Rather, we suggest creating a series of aggressive, de-centralized launches in several targeted local and regional markets across the country. This approach: . . .

- Avoids cynical reporters from major media: less reviewing/challenging of TASSC messages.[68]

The media coverage, the public relations company hoped, would enable TASSC to 'establish an image of a national grassroots coalition'.[69]

In case the media asked hostile questions, APCO circulated a sheet of answers, drafted by Philip Morris.[70] The first question was:

Isn't it true that Philip Morris created TASSC to act as a front group for it?

A: No, not at all. As a large corporation, PM belongs to many national, regional, and state business, public policy, and legislative organizations. PM has contributed to TASSC, as we have with various groups and corporations across the country.[71]

The fifth question was:

What areas of public policy would you like to see TASSC investigate?

A: We are not in a position to suggest that TASSC examine any issue; it's an independent organization and will no doubt proceed as best they determine.[72]

It should already have become clear that there are similarities between the language used and the approaches adopted by Philip Morris and by the organizations funded by Exxon. The two lobbies use the same terms, which appear to have been invented by Philip Morris's consultants. 'Junk science' meant peer-reviewed studies showing that smoking was linked to cancer and other diseases. 'Sound science' meant studies sponsored by the tobacco industry suggesting that the link was inconclusive.[73] Both lobbies recognized that their best chance of avoiding regulation was to challenge the scientific consensus. As a memo from the tobacco company Brown and Williamson noted,

Doubt is our product since it is the best means of competing with the 'body of fact' that exists in the mind of the general public. It is also the means of establishing a controversy.[74]

Both industries also sought to distance themselves from their own campaigns, creating the impression that they were spontaneous movements of professionals or ordinary citizens: the 'grassroots'.

But the connection goes much further than that. TASSC, the 'coalition' created by Philip Morris, was the first and most important of the corporate-funded organizations denying that climate change is taking place. It has done more damage to the campaign to halt it than any other body.

TASSC did as its founders at APCO suggested, and sought funding from other sources. Between 2000 and 2002 it received $30,000 from Exxon.[75] The website it has financed[76] – www.JunkScience.com – has been the main entrepot for almost every kind of climate change denial that has found its way into the mainstream press. While Fred Singer was the first to have posted the glacier figures on the web, it was JunkScience.com that popularized them. In fact you can still find them there today.[77] It equates environmentalists with Nazis, communists and terrorists. It flings at us the accusations which could justifiably be levelled against itself: the website claims, for example, that it is campaigning against 'faulty scientific data and analysis used to advance special and, often, hidden agendas'.[78] I have lost count of the number of correspondents who, while questioning manmade global warming, have pointed me there.

The man who runs it is called Steve Milloy. In 1992, he started working for APCO – Philip Morris's consultants.[79] While he was there, he set up the JunkScience site.[80,81] In March 1997, the documents show, he was appointed TASSC's executive director.[82] By 1998, as he explained in a memo to the board members, his JunkScience website was was being funded by TASSC.[83] Both he and the 'coalition' continued to receive money from Philip Morris. An internal document dated February 1998 reveals that TASSC took $200,000 from the tobacco company in 1997.[84] Philip Morris's 2001 budget document records a payment to Steven Milloy of $90,000.[85] Altria, Philip Morris's parent company, admits that Milloy was under contract to the tobacco firm until at least the end of 2005.[86]

He has done well. You can find his name attached to letters and articles seeking to discredit passive-smoking studies all over the internet and in the academic databases. He has even managed to reach the *British Medical Journal*: I found a letter from him there which claimed that the studies it had reported 'do not bear out the hypothesis that maternal smoking/passive smoking increases cancer risk among infants'.[87] TASSC paid him $126,000 in 2004 for fifteen hours of work a week.[88] Two other organizations are registered at his address: the Free Enterprise Education Institute and the Free Enterprise Action Institute.[89] They have received $10,000 and $50,000 respectively from Exxon.[90] The secretary of the Free Enterprise Action Institute is a man called Thomas Borelli.[91] Borelli was the Philip Morris executive who oversaw the payments to TASSC.[92]

Milloy also writes a weekly 'Junk Science' column for Fox News. Without declaring his interests, he has used this column to pour scorn on studies documenting the medical effects of second-hand tobacco smoke and showing that climate change is taking place.[93,94,95,96,97] Even after Fox News was told about the money he had been receiving from Philip Morris and Exxon,[98] it continued to employ him, without informing its readers about his interests. It still describes him thus:

Steven Milloy publishes JunkScience.com, CSRWatch.com. He is a junk-science expert, an advocate of free enterprise and an adjunct scholar at the Competitive Enterprise Institute.[99]

TASSC's headed notepaper names an advisory board of eight people.[100] Three of them are listed by Exxonsecrets.org as working for organizations taking money from Exxon. One of them is Frederick Seitz, the man who wrote the Oregon Petition, and who chairs Fred Singer's Science and Environmental Policy Project.

In 1979, Seitz became a permanent consultant to the tobacco company RJ Reynolds.[101] He worked for the firm until at least 1987,[102] for an annual fee of $65,000.[103] He was in charge of deciding which medical research projects the company should fund,[104] and handed out millions of dollars a year to American universities.[105] The purpose of this funding, a memo from the chairman of RJ Reynolds shows, was to 'refute the criticisms against cigarettes'.[106] An undated note in the Philip Morris archive shows that it was planning a 'Seitz symposium' with the help of TASSC, in which Frederick Seitz would speak to '40–60 regulators'.[107,108]

S. Fred Singer also had connections with the tobacco industry. In March 1993, APCO sent a memo to Ellen Merlo, the vice president of Philip Morris who had just commissioned it to fight the Environmental Protection Agency.

As you know, we have been working with Dr. Fred Singer and Dr. Dwight Lee, who have authored articles on junk science and indoor air quality (IAQ) respectively. Attached you will find copies of the junk science and IAQ articles which have been approved by Drs Singer and Lee. . . . We discussed with Dr Singer Ellen's suggestion for the junk science article to have a more personal introduction, however he is adamant that this would not be his style. Please review the articles and let us know as soon as possible whether you have any comments or questions about them.[109]

Singer's article, entitled *Junk Science at the EPA*, claimed that

The latest 'crisis' – environmental tobacco smoke – has been widely criticized as the most shocking distortion of scientific evidence yet.'[110]

He alleged that the Environmental Protection Agency had had to 'rig the numbers' in its report on passive smoking. This was the report that Philip Morris and APCO had set out to discredit a month before Singer wrote his article.

In another note, APCO reveals that it has discussed with Fred Singer the means of organizing an international movement to support TASSC's aims.[111]

I have no evidence that Fred Singer or his organization have taken money from Philip Morris. But many of the other bodies which have been sponsored by Exxon and have sought to repudiate climate change were also funded by the tobacco company.[112] Among them are some of the world's best-known 'think tanks': the Competitive Enterprise Institute, the Cato Institute, the Heritage Foundation, the Hudson Institute, the Frontiers of Freedom Institute, the Reason

Foundation and the Independent Institute, as well as George Mason University's Law and Economics Center.[113] I can't help wondering whether there is any aspect of 'conservative' thought in the United States which has not been formed and funded by the corporations.

Until I came across this material, I believed that the accusations, the insults and the taunts such people had slung at us environmentalists were personal: that they really did hate us, and had found someone who would pay to help them express those feelings. Now I realise that they have simply transferred their skills.

But they are taken seriously throughout the English-speaking media. This is how the BBC introduced an online debate it hosted in July 2004:

Ask the experts: What does the future hold for climate change? ... Your questions were answered by former Environment Minister Michael Meacher and global climate-change expert Dr S. Fred Singer.[114]

The BBC's debate pitches a non-scientist against a scientist: the authority, in other words, appears to lie with the 'global climate-change expert' Dr Singer, though his publication record over the past twenty years has been sparse. The BBC chose him rather than any of the thousands of environmental scientists with stronger credentials because its editors believed that this was where the debate lay: between those who claimed climate change was happening, and those who claimed it was not. They believed it, despite several reminders to the contrary from the Royal Society, because Fred Singer and Steve Milloy and others had created this impression through their appearances in the media. The story was self-perpetuating.

Until mid 2005, the BBC seemed incapable of hosting a discussion on climate change without bringing in one of the Exxon-sponsored deniers to claim that it was not taking place. On only one occasion did it tell its listeners that the 'expert' it had chosen had been funded by an oil company.[115] It could be argued that by failing to declare their interests it was providing free, unacknowledged airtime for the corporation. It now appears to have woken up to the extent to which it has been (in the words of one senior executive I spoke to) 'fooled by those people'. But in the United States and Australia, Exxon's experts are still being presented as serious scientists. Steve Milloy, for example, has appeared on CNN, ABC, MSNBC, National Public Radio and most of the major programmes on the Fox News network.[116]

They are also taken seriously by politicians. In 2003, James Inhofe, the Republican senator from Oklahoma, delivered a speech to the Senate called 'The Science of Climate Change'. Here is an extract.

The claim that global warming is caused by manmade emissions is simply untrue and not based on sound science.

Carbon dioxide does not cause catastrophic disasters – actually it would be beneficial to our environment and our economy.

. . . With all of the hysteria, all of the fear, all of the phoney science, could it be that manmade global warming is the greatest hoax ever perpetrated on the American people? It sure sounds like it.[117]

How did he know? Because he had spoken to 'the nation's top climate scientists', whom he then proceeded to list. The list began with Dr S. Fred Singer. It went on to name Frederick Seitz, the two employees of the George C. Marshall Institute who wrote the 'review' Seitz had circulated, and eight others working with organizations sponsored by Exxon.[118] Alarmingly, Inhofe continues to chair the Senate Committee on Environment and Public Works.

In 2004, *Harper's* magazine published a leaked memo from Myron Ebell of the Competitive Enterprise Institute to Phil Cooney, the chief of staff of the White House Council on Environmental Quality. The Competitive Enterprise Institute has been given over $2m by Exxon.[119] In 1997, the only year for which I have records, it received $125,000 from Philip Morris.[120] Ebell's memo showed that the White House and the Institute had been working together to discredit a report on climate change produced by the Environmental Protection Agency, whose head at the time was Christine Todd Whitman.

Dear Phil,

Thanks for calling and asking for our help . . . As I said, we made the decision this morning to do as much as we could to deflect criticism by blaming EPA for freelancing. It seems to me that the folks at EPA are the obvious fall guys, and we would only hope that the fall guy (or gal) should be as high up as possible. I have done several interviews and have stressed that the President needs to get everyone rowing in the same direction. Perhaps tomorrow we will call for Whitman to be fired.[121]

The New York Times later discovered that Phil Cooney, who is a lawyer with no scientific training, had been imported into the White House from the American Petroleum Institute, to control the presentation of climate science.[122] He edited scientific reports, striking out evidence that glaciers were retreating and inserting phrases suggesting that there was serious scientific doubt about global warming.[123] When the revelations were published he resigned and took up a post at Exxon.[124]

The oil company also has direct access to the White House. On 6 February 2001, seventeen days after George W. Bush was sworn in, A. G. (Randy) Randol, ExxonMobil's senior environmental adviser, sent a fax to John Howard, an environmental official at the White House.[125] It began by discussing the role of Robert Watson, the head of the Intergovernmental Panel on Climate Change. It suggested he had a 'personal agenda' and asked

Can Watson be replaced now at the request of the US?[126]

It went on to ask that the United States be represented at the panel's discussions by a Dr Harlan Watson.[127] Both requests were met. One Watson was sacked, the other was appointed, and he continues to wreak havoc at international climate meetings.

While they have been most effective in the United States of America, the impacts of the climate-change deniers sponsored by Exxon and Philip Morris have been felt all over the world – I have seen their arguments endlessly repeated in Australia, Canada, India, Russia and the United Kingdom. By dominating the media debate on climate change during seven or eight critical years in which urgent international talks should have been taking place, by constantly seeding doubt about the science just as it should have been most persuasive, they have justified the money their sponsors spent on them many times over. I think it is fair to say that the professional denial industry has delayed effective global action on climate change by several years.

*

None of this is to suggest that the science should not be subject to constant scepticism and review, or that environmentalists should not be held to account. It is only through repeated challenges to accepted wisdom that science has progressed. Climate-change campaigners have no greater right to be wrong than anyone else: if we mislead the public, we should expect to be exposed. We also need to know that we are not wasting our time: there is no point in devoting your life to fighting a problem that does not exist.

But the people who have been paid by Exxon are not, as they claim, 'climate sceptics'. They do not fit the usual definition of 'sceptic':

A seeker after truth; an inquirer who has not yet arrived at definite conclusions.[128]

They are members of a public relations industry, which begins with a conclusion and then devises arguments to support it.

Nor is this to suggest that political resistance to dealing with climate change is entirely the work of these people. The US government, for example, doesn't need Exxon's help to sabotage the international climate negotiations. One of the reasons why the professional climate-change deniers have been so successful in penetrating the media is that the story they have to tell is one that people want to hear.

In one respect, almost everyone – including the campaigners – is in denial about climate change. We have chosen to believe that the targets set by some of the more progressive governments provide realistic means of dealing with it. The United Kingdom, for example, intends to cut carbon emissions by 60 per cent by 2050. This is one of the world's most ambitious objectives. It is also, as I hope the previous chapter demonstrated, next to useless. But most of the climate-change reports produced by the major environmental groups seek to demonstrate that this target can be met without major economic loss. Whether or not it can be met is irrelevant: it is the wrong target. None of them has yet stepped forward to say that we need a cut of the magnitude the science demands.

The British government's chief scientist, Sir David King, has been heroic in drawing attention to the dangers of climate change, and he has taken plenty of flak for it. In a speech in October 2004, he said the following:

So what is the point at which the Greenland ice sheet will start melting? The latest indication is when the temperature around the Greenland land mass area is 2.7° centigrade above the pre-industrial level . . . What level of carbon dioxide therefore do we need to avoid going beyond in order to avoid testing this theory of melting of the Greenland ice sheet? . . . I used to say 550 parts per million, I'm now thinking that that might be pushing it a bit. So we are talking about perhaps 500 parts per million as the level beyond which we shouldn't go.[129]

In September 2005, I attended a conference in London at which Sir David was speaking. He told it that a 'reasonable' target for stabilizing carbon dioxide in the atmosphere was 550 parts per million. This happens to be the target set by the British government. It would be 'politically unrealistic', he said, to demand anything lower.[130] Simon Retallack from the Institute for Public Policy Research stood up and reminded Sir David that as chief scientist his duty is not to represent political reality, but to represent scientific reality. Retallack's own work shows that at 550 parts per million the chances of preventing more than 2° of global warming are just 10–20 per cent.[131] Sir David replied that if he recommended a lower limit, he would lose credibility with the government.

I think many people feel like him: that if they adopted the position determined by science rather than the position determined by politics, no one would take them seriously.

But the thought that worries me most is this. As people in the rich countries – even the professional classes – begin to wake up to what the science is saying, climate-change denial will look as stupid as Holocaust denial, or the insistence that AIDS can be cured with beetroot. But our response will be to demand that the government acts, while hoping that it doesn't. We will wish our governments to pretend to act. We get the moral satisfaction of saying what we know to be right, without the discomfort of doing it.

My fear is that the political parties in most rich nations have already recognized this. They know that we want tough targets, but that we also want those targets to be missed. They know that we will grumble about their failure to curb climate change, but that we will not take to the streets. They know that nobody ever rioted for austerity.

This is a gloomy thought. But it does reinforce my belief that we must make the necessary changes as painless as possible.

3

A Ration of Freedom

If thou repent, devils shall tear thee in pieces.
Doctor Faustus, Act II, Scene 3[1]

Making the necessary changes as painless as possible means, among other measures, making them as fair as possible. Even if we put the moral considerations aside, there is a good political reason for fairness. People are more willing to act if they perceive that everyone else is acting. In his book *Happiness: Lessons from a New Science*, Professor Richard Layard observes that

The only situation where we might willingly accept a pay cut is when others are doing the same. That is why there was so little economic discontent during the Second World War.[2]

George Orwell made a similar point in his essay *The Lion and the Unicorn*, published in December 1940.

The lady in the Rolls-Royce car is more damaging to morale than a fleet of Goering's bombing planes.[3]

In other words, one of the criteria for success in cutting greenhouse gas emissions by 90 per cent is that the cut applies to everyone. This would ensure that I was no longer confronted with the utter futility of buying an energy efficient bulb while Mr Meikle from Coalburn attaches a million Christmas lights to his house.

If we attempted to suppress climate change entirely by means of energy taxes, two things would happen. The poor would be hit much harder than the rich, as the costs took up a higher proportion of their income. And the rich would be able to carry on burning as much fuel as they could afford. In theory, you could design a tax and rebate system which ensured that money was transferred from the rich to the poor and which was constantly adjusted to maintain a steady cap on the amount of carbon the country produced. But while it would be no harder to implement than a rationing system, it would, because of the complex system of fees and rebates, be more difficult to explain. Complex ideas seldom do well in politics, as most people do not have the time or patience required to understand them. We are likely to react against one part of the package before we have grasped the whole idea.

An alternative approach is to draft new laws governing every move we make, before which, in theory, all men and women would be equal. We could have regulations governing when we could turn the lights on or how far we were allowed to travel. I don't believe that many people would see that as an attractive option either. As the Allied powers' economic planners found in the Second World War, there is a less coercive system, whose fairness is immediately apparent. It is rationing.

Rationing begins with a decision about the amount of carbon the world can emit every year. If, for example, it is correct to say that our 7 billion tonnes of current carbon emissions must be reduced to 2.7 by 2030, and if we want to make the biggest cuts sooner rather than later, we might decide that in 2012 the world should be producing no more than 5.5 billion tonnes. We divide that figure by the number of people we will expect to find on earth in 2012, and discover how much carbon everyone would be entitled to emit: it would be around 0.8 tonnes. Every nation would then multiply that figure by the number of people it contained, and this would become its national allocation.

This means that some countries, generally the poorest ones, would be allowed to raise their emissions: even in 2030, Ethiopia, if its population remained stable, could emit five and a half times as much carbon as it does today. But the overall effect would be an annual contraction of global carbon emissions, as the different countries converged towards the same amount per person. Unsurprisingly, this approach is known as 'contraction and convergence'.[4] It was devised by a man called Aubrey Meyer. He is one of those extraordinary people whose lack of relevant qualifications appears to work in his favour: he's a concert viola player. Meyer was able to leap over the more constrained proposals of the professionals and produce an idea that was simple, based on science and fair.

Once a country has its allocation, it can then decide how its emissions should be parcelled out. In theory, you could simply hand everyone his or her global share: 0.8 tonnes of carbon, for example. But this, though at first it seems straightforward, would create an incredibly complex system. Everything you bought would need both a cash price and a carbon price. If, for example, you stopped beside the road to buy a punnet of strawberries, you would need to pay, say, £1 for it, plus 0.01588 per cent of your carbon entitlement – assuming that someone had worked out that the growing, transport and packaging of the strawberries had caused 127 grams of carbon to be released. It's not going to work.

A much simpler system was devised by Mayer Hillman and refined by another independent thinker, a man called David Fleming. Both companies and people would need to use their carbon accounts when buying just two commodities: fuel and electricity.[5] If, for example, the fuel and electricity that people consumed directly added up to 40 per cent of a country's carbon emissions, then the citizens of that country would be given 40 per cent of its carbon budget. Everyone would get the same amount, and no one would have to pay. We would need to use our carbon allowance only when paying our electricity or gas bills or filling up our cars. (Fleming's scheme could be extended a little, to cover aeroplane and train journeys as well.[6])

If, then, the global carbon allocation in 2012 is 0.80 tonnes (800 kilograms) – of carbon per person, each of us would be given a carbon debit card entitling us to spend 40 per cent of that, or 320 kilograms.

The remaining 60 per cent of the country's carbon budget would belong to the government. It keeps some for itself and auctions the rest either directly to companies wanting to buy fuel or electricity, or to carbon brokers, who would then sell the entitlements to other corporations or to people who cannot stay within their budgets. The price, like that of any other commodity, would depend on the competition for the resource, which in turn would depend on its scarcity. So by the time you stop to buy the punnet of strawberries, the carbon required to produce it would already have been incorporated into its price, and you need to pay only in pounds. The more carbon-intensive a product is, the more expensive it will be.

But in creating a carbon rationing system, you are, in effect, creating a new currency. The entitlement to pollute will be accounted, saved, spent and exchanged much as money is today. As far as I can discover, no one has yet given it a name, except the rather dull 'carbon units'. So for want of a better term, I will call the new currency icecaps, in the hope that the name will remind people what the system is for: it enables us to cap our carbon emissions to keep the planet cool.

The icecaps you are given can be traded with other people. If you reach the end of the year and find you haven't used all your allocation, you can sell the remainder to someone else. Or if you've used too much, you can buy the extra icecaps you need. You can buy or sell the unused rations in your local post office or bank, or from electricity companies, filling stations and travel agents. So can visitors to the country, who will not have been given any entitlements by the government.[7] Of course, if everyone is trying to do the same thing, the price becomes very high indeed.

What this means is that the lady in the Rolls-Royce car might still be driving around, but only after she has transferred a good deal of money to people who are poorer or more abstemious than she is. Economic justice is built into the system.

It is a much fairer plan than, for example, the European Union's Emissions Trading Scheme. This system, which has been running since the beginning of 2005, began by handing out carbon dioxide emissions permits, free of charge, to big European companies. By and large, those who produced the most carbon emissions were given the most permits: the polluter was paid.[8] This handout was so generous that, in May 2006, the British government's consultants calculated that power firms would be making a windfall profit from the scheme of around £1 billion, while doing nothing to reduce their emissions.[9] The Emissions Trading Scheme is a classic act of enclosure. It has seized something which should belong to all of us – the right, within the system, to produce a certain amount of carbon dioxide – and given it to the corporations.

While rationing is, of course, a matter of deprivation and restraint, it offers greater freedom than the competing means of reducing carbon emissions. If, as I have suggested, the government sought to hold our emissions down to the necessary level by means of separate regulations, it would have to police our behaviour relentlessly. We would need to be punished for failing to turn our televisions off at the wall, or for leaving the lights on when we go to bed. Some technologies would have to be banned outright. But under a rationing scheme, as long as you either stay within your allocation or are prepared to buy extra icecaps, you can do whatever you want. If you can afford it, you can burn your entire ration in a single carbon orgy, then buy what you need for the rest of the year from other people. You can watch television all day long, and make up for it by swapping your fridge for a super-efficient model. It doesn't matter. What counts is that the country as a whole will not be exceeding its share of carbon dioxide.

The market created by carbon rationing will automatically stimulate demand for low-carbon technologies, such as public transport and renewable energy. In other words, in every respect this proposal needs be less statist than its competitors.

It is not quite as simple as this, however. While in general there is a direct relationship between wealth and energy use, it doesn't hold all the way across society. In the United Kingdom, for example, 30 per cent of the very poor (those who live in the poorest one fifth of households) use more energy than the national average.[10] The main reason for this is that they live in terrible houses. Some people have to burn so much fuel to keep warm that they would scarcely be worse off if they lived in the open air. A carbon rationing scheme, in other words, cannot be just unless it is accompanied by a massively accelerated programme to improve the condition of the poorest people's homes.

The poor might also be living in places which are badly served by public transport: they need taxis or old bangers in order to get to work or to the shops. Again, help is needed – a better public transport system, for example – if the poorest people are not to be faced with a choice between food and energy. In other words, simply creating a market and expecting it to solve the entire greenhouse gas problem is like asking the people of the slums of Manchester in the 1840s to sort out their own sanitation. There were some things they could do, if they could afford them: boiling their water, washing with soap, changing their clothes more often. But they could not build sewerage and clean water systems by themselves and – because they were poor – no one had a financial interest in providing them. The massive investment in new services which carbon rationing demands must, in some places, be assisted by either funding or guidance from the state.

The government might also need to hold back part of the national allowance as an extra ration for people on low incomes who find themselves in trouble during a hard winter. It could function much as cold weather payments do in northern European countries today: when the temperature drops below a certain point, governments give an allowance to the poorest people to help them pay their fuel bills.[11]

Of course, the more energy the carbon ration allows us to use, the more politically acceptable the scheme becomes. If a 90 per cent carbon cut means a 90 per cent energy cut, it is hard to see how any democratic government could make it happen. We must seek, as far as possible, to break the link between carbon and energy. If, for example, the entire electricity network were somehow to be supplied by renewable power, then we wouldn't need to hand over any icecaps to keep the lights on. The purpose of this book is to find the most politically effective means both of cutting our energy use and of reducing its carbon content.

None of this automatically solves the political problem. As every government strives to minimize the restraint upon the lives of its citizens, international climate negotiations have so far been detached from what the science is saying. There is still no global target for greenhouse gas concentrations in the atmosphere. The cuts agreed under the Kyoto Protocol – 5.2 per cent by 2012 – bear no relationship to the cuts the problem demands. The Kyoto Protocol resembles the European Emissions Trading Scheme: a country's entitlement to pollute relates to the amount of pollution it already produces. The dirtier you are, the bigger your entitlement.

But while adopting the principle of contraction and convergence would not mean an end to the political arguments, they would no longer take place in a moral and intellectual vacuum. The negotiators would have a target – an equal division of the planet's capacity to absorb pollution – which is both factual and fair. The best estimate of the planet's total carbon sink in 2030 will change as the science improves, but the target can change with it. With an equal global carbon allocation, countries will no longer be able to claim that they can't act because others are not obliged to join in. They might not like this proposal, but they cannot deny that it is even-handed.

The argument deployed most often against a global carbon reduction of this kind, whether it is brokered by rationing or any other system, is that the cure is worse than the disease. Just as Dr Faustus was told that if he repented, devils would tear him apart, we are told that the economic decline it would cause would outweigh any losses imposed by climate change.

This position has been most famously advanced by the Danish statistician Bjørn Lomborg, whose work delights the media on both sides of the Atlantic. In his book *The Skeptical Environmentalist*, which is thorough and well-referenced, Lomborg argues that the cost of doing nothing about climate change – of letting nature take its course – is US $4,820 billion. The cost of stabilizing global temperatures at 2.5° above the 1990 level would be $8,553 billion; and the cost of stabilising them at 1.5° above the 1990 level (roughly what I am seeking to achieve here) would be $37,632 billion – nearly 38 trillion dollars.[12] He warns that

With the best intentions of doing something about global warming we could end up burdening the global community with a cost much higher or even twice that of global warming alone.

It would be better to use our money to make investments with higher returns, leaving 'future generations of poor people with far greater resources'.[13]

I could seek to counter Lomborg's case, as many environmental economists have done, by arguing that he and the economists upon whose figures he relies are wrong: that the economic costs of letting climate change happen greatly outweigh the economic costs of tackling it. But I will not do so for this reason: it is an amoral means of comparison.

We can determine, for example, that the financial costs of Hurricane Katrina, which may have been exacerbated by climate change, amount to some $75 billion,[14] and we can use that number to help derive a price for carbon pollution. But does it capture the suffering of the people whose homes were destroyed? Does it capture the partial destruction, in New Orleans, of one of the quirkiest and most creative communities on earth? Does it, most importantly, capture the value of the lives of those who drowned?

In other words, is it possible to place an economic price on human life? Or on an ecosystem, or on the climate? Could such costs, when rolled out around the world, really be deemed to amount to $4,820 billion, give or take the odd dollar? If you believe the answer is yes, then I charge that you have spent too much time with your calculator and not enough with human beings.

When economists have tried to cost such things, they have simply exposed the limitations of their science. In 1996, for example, a study for the Intergovernmental Panel on Climate Change estimated that a life lost in the poor nations could be priced at $150,000, while a life lost in the rich nations could be assessed at $1.5 million.[15] The researchers produced these figures by estimating how much people would be prepared to pay for the adaptive measures that would save their lives. Unsurprisingly, they discovered that the lives of rich people were worth more than the lives of poor people. Their figures were not just wrong; they were meaningless.

These days, economists are less prepared to expose themselves to ridicule. So anything that cannot be quantified is simply excluded from the balance sheet. What this means is that the loss of all the really important things – a functioning ecosystem, human communities, human life – is overlooked. Because they aren't counted, they don't count.

It would be wrong to blame only Bjørn Lomborg and the economists whose work he promotes for taking this line. Even those who claim to want to do something about climate change can't help themselves: they feel obliged to attach a price tag to global catastrophe. The British government, for example, has decided that the 'social cost' of carbon emissions is somewhere between £35 and £140 per tonne, with a middle value of £70.[16] We might reasonably ask what the heck this means. Does the government really believe it can put a price on the Amazon? On Bangladesh? This must, in other words, be a moral decision, not an economic one. Either we decide that it is right to spend a lot of money seeking to prevent catastrophic climate change or we decide that it isn't, but we must make that decision on the grounds of how much we value people and places as people and places, rather than as figures in a ledger.

This does not relieve me, however, of the need to work out how much the proposals in this book might cost.

When I say that this isn't easy, it would be an understatement to call it an understatement. I haven't come across such wildly varying claims in any other field. Bjørn Lomborg's extraordinarily precise figure – $37.632 trillion – occupies one extreme; at the other are people who claim that cutting carbon emissions will actually *make* us money, as the requirement to invest in new technologies will stimulate economic growth, and energy efficiency will lead to financial efficiency. The British government, for example, estimates that saving 37 per cent of the energy currently used in our homes would also save around £5 billion.[17]

In between these extremes, you can take your pick. The European Commission chooses studies suggesting that stabilising carbon dioxide emissions at 550 parts per million by 2100 would cost the world somewhere between $1 trillion and $8 trillion.[18] Keeping them below 450 parts per million would cost between $2.5 trillion and $18 trillion. The economist Christian Azar and the biologist Stephen Schneider have produced a similar figure for a completely different target: stabilizing carbon dioxide emissions at 350 parts would cost about $18 trillion.*[19]

These are all stupendous amounts of money. My first reaction on encountering them was that, though they were generally lower than Bjørn Lomborg's costs, they were still completely unaffordable. But as Azar and Schneider point out, this would be true only if we had to pay them all at once. If we spread the cost between now and the target date, it would hurt us much less. The modellers who have

---

* All these figures refer to carbon dioxide only. If you included the other greenhouse gases, the equivalent concentrations would be about 15 per cent higher.

frightened us with such huge sums also predict that the global economy will keep growing by 2–3 per cent a year through the twenty-first century, which means it will be in the order of ten times bigger in 2100 than it was in 2000. Holding down carbon emissions at a cost of trillions or even tens of trillions of dollars would merely delay the point at which we became ten times richer than we are today by a few years. 'To be ten times richer in 2100 AD versus 2102 AD,' Azar and Schneider point out, 'would hardly be noticed.'[20] My target date (2030) is much closer than theirs, however, so the economic impacts would be felt more keenly.

I could claim that my proposals would cost a total of $10 trillion, or 20 or 30, or I could create the impression that I knew exactly what was going to happen, by telling you that they will cost $14.739 trillion. But you could trust any one of those estimates only as far as you could trust all the assumptions – about economic growth, discount rates, energy prices, new technologies, government policy – that lay behind it. Because the assumptions range so widely as to render each other almost meaningless, any single figure I offered would be an arbitrary one. But what I can do is to try to give you some idea of the economic hardship which might be caused.

One means of doing this is to compare the way prices might be expected to change under the proposals in this book with price fluctuations in the past. Again, as Chapters 5 to 7 will show, this is difficult, as the estimates of the costs of energy produced by different means also vary widely. But roughly speaking, my proposals are likely to raise the wholesale price of electricity and heat, in real terms, by something in the order of 100 per cent by 2030. This sounds like a lot, until you see it in context. Between November 2004 and November 2005, the average wholesale price of electricity rose from 2.1 pence to 3.6 pence – by 71 per cent. In the twelve months to February 2006, the wholesale price of natural gas in the United Kingdom rose by 75 per cent.[21] In the three years to that date, it rose from under 20p a therm to 70p – an increase of 350 per cent.[22] The rate by which the price of gas has risen is thus roughly twenty-eight times greater than the rate of increase in energy prices which would be caused by my proposals. The new gas prices have caused some pain, but they have not led to economic collapse.

The greater impact is one that will not really be 'felt'. It is the effect upon what we might have been able to do had the restraints I am advocating not been in place. The money which needs to be spent to tackle climate change and the economic opportunities which have to be foregone are regularly confused – sometimes, I think, deliberately – by those who wish to prevent us from taking action. At the moment, all of us suffer from the economic constraints imposed by the fact that we are not allowed to buy psychotropic drugs. If this trade were legal, it would have enhanced the pace of economic growth, which means that on average we would be slightly richer today than we are. But, with the exception of a few people who have been hoping to sell these items without fear of prosecution, we do not walk around under a cloud, lamenting the fact that money which might have entered our pockets has been denied to us. The impact of this constraint is real, but it has not been felt by us. No one's bank account is emptied by what might have been.

Having taken the plunge into comparative economic pain, the system I propose has one self-limiting characteristic. There is a pay-off between the price of energy and the price of icecaps. If we invest heavily in renewable electricity, for example, and therefore weaken the link between energy and carbon, every kilowatt hour of electricity will cost more. This is because – at the moment – most renewables are more expensive than coal or gas. On the other hand, we can consume more electricity while staying within our carbon ration. As fewer people will then need to buy extra rations, the price of icecaps will be suppressed. As this lower price feeds through the economy, it counteracts the inflation caused by higher electricity prices. The over-all cost of energy would be higher than it is today, but to some extent it holds itself down.

None of this means, however, that I can give myself a free hand in suggesting how climate change might be tackled. In every case I must seek the cheapest way to cut carbon emissions, for the reason that a pound spent on an inefficient process is a pound not spent on an efficient one.

But there is a further question we should ask. If these changes are to cost money, on whom or on what will the cost be imposed?

One of the arguments made by those who claim that we should

take no action is that if the same amount of money were spent on relieving hunger, or supplying clean water, or preventing AIDS or tuberculosis or malaria, it would save more lives.[23] This approach tends to overlook the fact that climate change is likely to cause more hunger, more water stress and more communicable disease, thereby raising the cost of addressing them. But this is not the principal argument against it. Behind this case is an unfathomable assumption: that money spent on preventing climate change is money not spent on foreign aid. In other words, it supposes that the climate-change budget is in direct competition with the rich countries' foreign aid budgets, rather than with any other kind of spending.

If the rich countries were already doing everything in their power to help the poor and were, as a result, now running out of money, this argument might carry some weight. But it is hard to think of a national exchequer which has ever been endangered by its foreign aid spending. The governments of Europe have agreed that by 2015 – a mere thirty-five years after the date they first set for themselves – they will give 0.7 per cent of their national income in foreign aid. The United States currently spends 0.17 per cent of gross domestic product – just over $19 billion – on aid.[24] The United Kingdom spends 0.36 per cent, or £4.3 billion.[25]

In other respects these governments are rather more open-handed. The United Kingdom's budget for the widening of the M1 motorway is £3.6 billion.[26] This is nearly seven times as much as it is currently spending every year on tackling climate change.* Altogether it has set aside £11.4 billion for building and widening roads. Given that the government has – or had – a policy of reducing road traffic, it is hard to see how this spending can be justified.

In his book *Perverse Subsidies*, published in 2001, Professor

---

* It has proved remarkably hard to obtain real figures from the UK government. In a parliamentary answer, the government spokesman Lord McKenzie claimed that 'at least £545 million is spent annually on spending policies that tackle climate change,'[27] but – though I have made three requests – the government has so far failed to give me a breakdown of this figure. The phrase 'spending policies that tackle climate change' makes me suspicious, as it could be used to include spending whose impact on climate change is incidental, or spending which tackles climate change among other objectives. Why did he not say simply 'at least £545 million is spent annually on tackling climate change'?

Norman Myers adds the direct payments US corporations receive from the government to the wider costs they oblige society to carry, and arrives at an annual figure of $2.6 trillion.[28] This is roughly five times as much as the profits they were making at the time his book was written. As well as the annual $362 billion the thirty richest governments were paying their farmers when *Perverse Subsidies* was published, they were spending some $71 billion on fossil fuels and nuclear power and a staggering $1.1 trillion on road transport. Worldwide, governments pay companies $25 billion a year to destroy the earth's fisheries, and $14 billion to wreck our forests.[29]

The Energy Policy Act the Bush administration pushed through Congress in 2005 handed a further $2.9 billion to the coal industry and $1.5 billion to oil and gas firms.[30] According to the Democratic congressman Henry Waxman, the oil subsidy 'was mysteriously inserted in the final energy legislation after the legislation was closed to further amendment'.[31] Most of the money, he discovered, would be administered by

a private consortium located in the district of Majority Leader Tom DeLay ... The leading contender for this contract appears to be the Research Partnership to Secure Energy for America (RPSEA) consortium ... Halliburton is a member of RPSEA and sits on the board, as does Marathon Oil Company ...

In short ... taxpayers will hire a private consortium controlled by the oil and gas industry to hand out over $1 billion to oil and gas companies. There is no conceivable rationale for this extraordinary largesse. The oil and gas industry is reporting record income and profits. According to one analyst, the net income of the top oil companies will total $230 billion in 2005.[32]

In the European Union, according to the European Environment Agency, direct and indirect subsidies for the coal industry amounted to €13 billion in 2001, and subsidies to the oil and gas industry to €8.7 billion.*[33] Germany made direct payments to the coal industry of over €4 billion that year, which equates to €82,000 for every job

---

* Direct subsidies are straight payments to the industry by the state. Indirect ones take the form of tax rebates and other concessions.

in the mines it saved.[34] The United Kingdom supports its oil and gas industry by holding down value added tax on its products, at a cost to the Treasury of about €1.4 billion a year.[35]

At the beginning of 2006, Joseph Stiglitz, the former chief economist at the World Bank, and Linda Bilmes, an economist at Harvard, calculated that, on 'very conservative' estimates, the war in Iraq had so far cost the United States between $1 and $2 trillion.[36]

When you remember that only a smallish proportion of the cost of dealing with climate change will be borne by governments, it becomes clear that this is not a choice between state spending on climate change or state spending on foreign aid and essential public services. It is a choice between state spending on climate change or state spending on coal, oil, roads, farm subsidies, environmental destruction and unprovoked wars. We would do well to ask why governments seem to find it so easy to raise the money required to wreck the biosphere, and so difficult to raise the money required to save it.

Most of the cost, though, will be carried not by states but by corporations and citizens. It means that some of the money we might otherwise have spent on other goods and services will have to be spent instead on more expensive energy. While the evidence is mixed, this is likely to counteract some of the increase in employment which would otherwise have happened as a result of greater economic growth.[37] This assumes that our governments are correct in predicting continued economic growth this century if we carry on as we are. But that assumption is beginning to look unsafe.

It is now becoming pretty clear that within the lifetimes even of the late middle-aged people alive today, global petroleum supplies will peak and then go into decline. There is a great deal of dispute about when, within the next thirty years, this will happen. At the time of writing, some analysts are claiming that it might already be happening, and will only become visible in retrospect, as oil prices start to rise almost exponentially.[38] Others, who appear to be just as well qualified, maintain that it is unlikely to happen until 2037.[39] But there is little dispute about the proposition that it will take place.

In 2005 the US Department of Energy released a report called

Peaking of World Oil Production: Impacts, Mitigation, & Risk Management.[40] The department's consultants, led by the energy analyst Robert L. Hirsch, concluded that

without timely mitigation, the economic, social, and political costs will be unprecedented.[41]

It is possible to reduce demand and to start developing alternatives, they said, but this would take '10–20 years'.

Waiting until world oil production peaks before taking crash program action leaves the world with a significant liquid fuel deficit for more than two decades ... the problem of the peaking of world conventional oil production is unlike any yet faced by modern industrial society.[42]

The International Energy Agency estimates that a $10 increase in the price of a barrel of oil causes the rich nations to lose 0.4 per cent of their gross domestic product.[43] Asia loses 0.8 per cent and sub-Saharan Africa over 3 per cent.[44] An oil peak could raise the price by $100, $200 or almost any amount of money. It could, as Hirsch suggests, precipitate the worst economic depression the modern world has ever known. Several of the measures I propose in this book are of the kind that we should be pursuing even if climate change were not happening, in order both to delay the peaking of world oil supplies and to reduce the economic impact when it happens. When the oil peak is taken into account, it seems to me that judicious spending on reducing climate change is at least as likely to prevent a recession as to cause one.

But even without the oil problem, there can still surely be only one answer to the question of whether preventing catastrophic climate change is worth the sacrifice of a proportion – even a large proportion – of our predicted economic growth over the twenty-four years between now and 2030. If we in the rich nations do not act to prevent it, we are likely, over that period, to have more money in our pockets than if we do. We could spend that money on – well, on what exactly? More cars, more flights, more Barbie dolls, more tiger prawns? More roads, more farm subsidies, more wars? In either case it is hard to see how these delights will compensate for the damage to our lives that climate change will cause. I could put it another way, but someone else already has.

For vain pleasure of twenty-four years hath Faustus lost eternal joy.[45]

4

Our Leaky Homes

Who was it built this house so ill
With shovel and with spade?   *Faust*, Part II, Act V[1]

Argue with the executives of any industry which is failing to cut its carbon emissions, and they will say the same thing: it doesn't matter, because they can pay other people to do it for them. This is how the European Emissions Trading Scheme works and, for that matter, how the rationing scheme I proposed in Chapter 3 will work. It is true that emissions trading is cheaper and more efficient than a system in which everyone has to achieve the same cut within their own businesses.

But while the delegation of responsibility works quite well – in theory at least – when the target for cutting emissions is unambitious, the bigger the cut becomes, the greater the number of industries which have to reduce their own pollution, rather than buy a clear conscience from somebody else. A 90 per cent cut across the economy means that every sector must cut its emissions by roughly that amount. If, for example, the carbon produced by land transport, which currently accounts for 22 per cent of our emissions, were to be reduced by only 50 per cent, emissions across the rest of the economy would have to be cut by 98.2 per cent. While I think I can show that 90 per cent is just within the realm of possibility, it is plain that 98.2 per cent lies well beyond it. If I am to make my proposals work, I must be able to demonstrate that a 90 per cent cut can be achieved not just in some sectors but in all those I cover.

I have made this task slightly easier for myself by examining only some parts of the economy. I have not tackled offices, the hospitality industry and public services, for example, because the solutions required in these sectors are quite similar to those explored in the following four chapters, and to sustain your interest I want to try to avoid repetition. I have examined the carbon-cutting potential of just a few of the many business sectors in the rich nations: if I tried to be comprehensive, this would be a very long book. But I have covered the activities responsible for over 60 per cent of our emissions, including some of those which are most difficult to redeem.

Perhaps my gravest omission is the military, whose use of fuel is – depending on how you look at the problem – either very hard to address or, on paper, very easy. The supersonic jet, for example, is possibly the most environmentally damaging technology ever developed. There will never be an eco-friendly F-35 Joint Strike Fighter. If planes like this continue to be deployed, they will destroy the climate as effectively as they destroy their targets. But I am among those who believe that our armed forces should be greatly reduced. The majority of them have no defensive purpose in the true sense of that term. Lacking convincing enemies – in the form of well-armed and aggressive neighbouring states – I believe that countries like the United Kingdom should confine their military operations overseas to peacekeeping: as much for the sake of the environment as for the sake of public finance and world peace. But I will excuse myself from further examination of this question on the grounds that it would require another book.

In this chapter I look at our houses and the extent to which their carbon emissions can be reduced by means of energy efficiency. I discover that the potential efficiencies are not as great as I had hoped, so in the following three chapters I explore the means by which the carbon content of the heat and electricity which our homes and the rest of the economy consume can be reduced. I seek to show that a combination of energy efficiency and new power-generating technologies will allow us to achieve a 90 per cent cut in emissions while keeping the lights and the central heating on. In Chapter 8 I look at how we might cut emissions from surface transport (on roads and railways) by 90 per cent, and in Chapter 9 I do the same for aviation. In Chapter 10 I explore the potential for a 90 per cent cut in two industries which produce a great deal of carbon dioxide: shops and cement. But I will begin by examining the peculiar problem of energy efficiency.

The commonest and most understandable mistake made by people engaging with the problem of greenhouse gas emissions is to assume that energy efficiency is the same as energy reduction. People imagine that if a piece of equipment uses 30 per cent less energy than the one it replaced, that 30 per cent has been saved. This was what I believed before I had the misfortune to encounter the Khazzoom–Brookes Postulate.

The postulate works like this. As efficiency improves, people or companies can use the same amount of energy to produce more services. This means that the cost of energy for any one service has fallen. This has two effects. The first is that money you would other-wise have spent on energy is released to spend on something else. The second is that as processes which use a lot of energy become more efficient, they look more financially attractive than they were before. So when you are deciding what to spend your extra money on, you will invest in more energy-intensive processes than you would otherwise have done. The extraordinary result is that, in a free market, energy efficiency could *increase* energy use.

It sounds ridiculous, and my instinct, when I first came across it, was to try to argue my way out. But the facts were not kind to me.

The postulate is named after two economists – Daniel Khazzoom and Len Brookes – who formed their theory in 1979 and 1980.[2] But the effect was first noticed long before then. In his book *The Coal Question*, published in 1865, Stanley Jevons showed that cutting the amount of coal used to produce a ton of iron by over two thirds 'was followed, in Scotland, by a tenfold increase in total consumption, between the years 1830 and 1863'.[3] Since Jevons's book was pub-lished, the world's energy efficiency has improved by around 1 per cent a year.[4] That shouldn't be surprising. The steam engine Thomas Newcomen built in 1712 had an energy efficiency of 0.5 per cent;[5] a good diesel engine today turns about 45 per cent of its fuel into useful work.[6] But throughout that period, with the exception of a couple of small declines when energy crises pushed the price up, the world's total consumption rose steadily. Between 1980 and 2002, energy use in the thirty richest countries rose by 23 per cent,[7] even while they exported their most energy-intensive industries to poorer nations. There is some evidence to suggest that this happened because the cost of energy per service fell.[8] The Khazzoom–Brookes Postulate appears to explain why the corporations, by pursuing their own cost-cutting interests, have not saved the planet.

I should point out that it remains, for now, only a 'postulate', and it is fiercely disputed by some energy experts. But I wish to try to make my proposals as watertight as possible, so I will assume that Khazzoom and Brookes got it right. If they are wrong, it does my proposals no harm. But if they are right and we ignore them, we are in danger of devising a scheme for reducing carbon emissions which does not work.

The postulate is similar to, but not quite the same as, the other great paradox of energy efficiency, which is known as 'the rebound effect'. While the Khazzoom–Brookes works on the economy as a whole (the macroeconomic level), the rebound effect operates within your own pocket (the microeconomic level). If you live in a well-insulated house, you need burn less gas to maintain a certain temperature. But as your gas bills are therefore lower, you will be tempted to turn the temperature up. Car engines are far more efficient than they used to be, but over the past twenty years their fuel consumption has scarcely declined. The driver's lower overall fuel costs permitted manufacturers to make cars bigger, heavier and faster and to make them do more: such as power steering, air conditioning and heating the windows.

The rebound effect isn't, on the whole, as strong as the proposed macroeconomic impact. While the Khazzoom–Brookes Postulate suggests that energy use increases as a result of efficiency, the rebound effect merely ensures that energy use does not decline as much as it would otherwise have done.[9] It is less controversial than the postulate.

These paradoxes are blissfully ignored by most environmentalists. In their book *Natural Capitalism*, for example, Paul Hawken and Amory and Hunter Lovins – who in other respects are innovative and forceful thinkers – set out to

dispel the long-held belief that core business values and environmental responsibility are incompatible or at odds.[10]

One of the examples they use to show how energy efficiency could be good for both business and the environment is the way air transport is run. At the moment, in order 'to monopolize gates and air traffic slots', many of the airlines force passengers to fly to big airport hubs, and change planes in order to reach their eventual destinations. But if they used 'smaller and more numerous planes that go directly from a departure city to a destination', then 'air travel would cost less, use less fuel, produce less total noise, and be about twice as fast point-to-point.'[11]

This, of course, is true. But if aviation is cheaper and quicker, more people will want to use it. The net effect is likely to be an increase in emissions. Indeed, they cite the good example of Southwest Airlines, which has increased its profits by turning customers around more quickly.[12] It has increased its profits because, as a result of its efficiencies, it now handles more customers. Somehow they contrive to overlook this consequence.

None of this is to suggest that energy efficiency should not be pursued. But what the paradoxes appear to show is that in the absence of proper government policies, it is not just a waste of time: it is counter-productive. In January 2006, for example, the governments of Australia, the United States, China, India, Japan and South Korea launched what they called 'the Asia-Pacific Partnership on Clean Development and Climate'. The partnership, which Australia and the US devised as an alternative to the Kyoto Protocol, differs from that agreement in that it sets no binding target for reducing carbon dioxide emissions. Instead it relies entirely on developing and sharing new technologies designed to save energy and carbon.[13] What the Khazzoom–Brookes Postulate suggests is that it cannot possibly work.

So here is another powerful argument for a rationing system. If efficiency is to work for us rather than against us, the amount of carbon the economy uses must be capped. And the only fair means of capping it is to give everyone an equal share. Only then does energy efficiency make sense.

When applied to houses, I believe I have discovered a third paradox: that regulation enhances the sum of human freedom. This was not a willing finding. I stumbled across it in the course of making another discovery: that my house is an ecological disaster.

In the two years since we bought it, I have slowly been finding out that there is scarcely a disease from which my house does not suffer. Neither the walls nor the floors are lagged, the windows rattle, there are gaps in the roof insulation nine inches wide, and the lights are embedded in the ceiling – which means that much of the electricity they use is employed to illuminate the underside of the floorboards.

The man we bought it from is a property developer. When he acquired it from the son of the old woman who died there, it was a ruin. He must have spent about £60,000 restoring it. Had he spent an extra £1,000, he would have cut my gas bills in half. Fitting the roof insulation properly would have cost him next to nothing. Solid wall insulation would have cost more, but part of the price could have been offset by using standard light fittings instead of the more expensive embedded ones. As he was ripping up the floors anyway, it would scarcely have hurt him to have rolled out a few strips of fibre.

But if we were to do what he should have done, we would need to gut the house all over again. The ceilings would have to come down, the floors would have to come up, the built-in shelves and cupboards would need to be ripped off the walls and we would have to decant ourselves into a rented home until the work was finished. It would cost us something like £20,000 to put right, and our efforts would add almost nothing to the value of the house. If we had it to spare, it would be better to pay someone to put a wind turbine on a mountain.

Ironically, we bought this house partly for environmental reasons: it is close to the town centre and well served by bicycle lanes and public transport, so we don't need a car. It has plenty of natural light, and it is 100 yards from the nearest allotments, which means that I can grow zero-carbon vegetables. But because the energy choices of the developer were unrestricted, our choice was constrained. In my city, where the oldest houses are closest to the centre, there are almost no energy-efficient homes whose location allows you to live a low-carbon life.

Because he was refurbishing this house, rather than building it from scratch, the property developer was subject to building regulations which were both sparse and weak.[14] Even those which did apply, as we have now discovered, were not enforced.

When a house is refitted to a high standard, the work should last for twenty or thirty years. During that time, with an average turnover in the United Kingdom of seven years,[15] three or four households will live there. In other words, stricter regulations would constrain one set of people to do the work properly, and release three or four sets of people from the burden of living in houses which cost a fortune to heat. Even within an otherwise lightly regulated system of the kind carbon rationing permits, strict building rules would lead to a net increase in human freedom.

But the government of my country, still digging its ideological hole, insists that tougher rules would be 'an unwarranted intervention in the market',[16] restricting people's choice of how they lived their lives. When the Minister for Housing and Planning, Yvette Cooper, was urged to introduce proper energy efficiency standards for the refur-bishment of houses, she said that it would amount to 'unnecessary gold plating'.[17] I remember that every time I read my gas bill.

It is partly because of this massive failure on the part of the state that our homes are responsible for such a high proportion of the energy we use. While the demand for energy in the United Kingdom rose by 7.3 per cent between 1990 and 2003, in our houses it rose by 19 per cent.[18] Altogether they're responsible for 31 per cent of the energy consumed here.[19] Of that, 82 per cent is used for space and water heating.[20] This has risen by 36 per cent since 1970.[21]

We use more energy to heat our homes partly because their average temperature increased, between 1991 and 2002, from 15.5° to 19°.[22] This is a good thing: many people, especially the elderly, have been living in homes so cold that they pose a danger to human life. But it should have been easy to achieve this while greatly reducing the amount of heating we use. In fact, as I will be showing later in the chapter, there are houses which maintain an average temperature higher than 19° without any heating whatever. But in this country our homes act as warm air tunnels: they keep us warm almost incidentally, as the heat pours past us and into the street.

There are 17 million homes with cavity walls in the United Kingdom, but only 6 million with cavity wall insulation.[23] Given that injecting mineral fibres between the bricks is so cheap that it pays for itself within two to five years,[24] the 65 per cent of homeowners who choose not to use it must either be so poor they have no capital to spend, so poorly informed that they have never heard of the process, aware that someone else (the tenant) is picking up the heating bill, or perversely attached to burning money. In 2002, 10 per cent of homes still had no insulation of any kind – wall, floor or roof.[25] This miserable circumstance gave rise to the best unintentional pun I have heard on the radio:

In the field of home insulation, Britain lags behind.

In 2004, the government said it was planning to oblige anyone extending a house to improve the heat-saving properties of the whole building.[26] The logic was pretty clear: a bigger house, all else being equal, will lose more heat. By improving the insulation in the rest of the house, you would compensate for the effect of the extension. But this, alongside several other progressive measures, was dropped at the last minute, when the new regulations were published in September 2005. As Andrew Warren, the head of the Association for the Conservation of Energy, remarked,

the Building Regulations will be late coming into force, and have been substantially and deliberately weakened.[27]

Even the rules governing new homes, whose enforcement would cause far fewer political headaches, are feebler than the government originally proposed. While they were supposed to cut energy use by 25 per cent, now – even if they are enforced – the best they will achieve is 18 per cent.[28] But both targets are pathetic.

Houses which meet the building codes in Norway and Sweden use around one quarter of the energy of houses meeting the standards in England and Wales.[29] In fact, the building regulations in Sweden were tougher in 1978 than they are in Britain today.[30] In Germany the air tightness standard – which determines how leaky a house is allowed to be – is three times as stringent as the standard in Britain.[31]

Even the feeble rules we do possess are mostly unenforced. The energy efficiency regulations were first introduced in 1985, but since then there has not been a single prosecution for non-compliance.[32] This is not because our builders never break the rules. A study by the Building Research Establishment found that 43 per cent of the new buildings it tested, which had received certificates saying that they complied with the regulations, should have been failed.[33] Professor David Strong, the head of the Establishment, observes that plenty of new homes have the requisite amount of insulation in their lofts, but quite often it is still tied up in bales, as the builders, knowing that no one would be checking, couldn't be bothered to roll it out.[34]

One of the reasons for this is that the government has allowed builders to turn to the private sector to get their certificates. Whereas in the past only local authorities would enforce the rules, now you can pay an 'approved inspector' to certify the house. The inspectors compete for business, and the judgements they must make on energy efficiency, Strong points out, 'are purely subjective'.[35] The inevitable result is that they don't want to be seen as too strict, or the builders will never use them again.[36] It's not only in the UK that this approach has proved to be a disaster. Even in Sweden there has been a decline in standards since the private sector started carrying out inspections.[37]

In fact it is hard to see what possible incentive builders have to comply with the energy efficiency rules. The chances of being found out are low, and when they are, they don't get prosecuted. While the buyers of new homes are insured against other regulations not being met, no cover is provided for a failure to meet the energy requirements, so the builders have nothing to fear from the insurers either.[38] It is cheaper to build houses badly than to build them well.

At the time of writing, the UK is being prosecuted by the European Union for failing to implement the new directive on the energy performance of buildings.[39] Among the other sensible ideas the directive contains is an energy label for houses: when you buy one you are supposed to be able to see how efficient it is. But we already have a system a bit like this,* which is meant to be enforced by the building regulations. A study by researchers at De Montfort University found that 98 per cent of builders are failing to provide housebuyers with the information to which they're entitled.[40]

The reason for all these failures is simple. David Strong says that the low standards are

---

* The SAP, or standard assessment procedure.

the result of very effective lobbying in the UK from organizations . . . that have no desire really to change working practices or the quality of buildings they are constructing.[41]

There are some good builders in this country, but the government has sided with the bad ones. In doing so, it makes good building almost impossible. This is because, while the building regulations are ineffective at setting minimum standards, they are very effective at setting maximum ones. No builder, unless the client asks for it, will build a house that is better than the regulations demand.

What makes all this so frustrating is that even as the government was deciding which standard to apply, new buildings were being constructed that demonstrate a staggering potential for reducing the use of energy. The prototype is something called the *Passivhaus* (passive house), which was first developed in Germany in the late 1980s.

There is no single design for a passivhaus: from the outside it can be almost indistinguishable from other modern houses. But there is something odd about them that you notice soon after stepping indoors: they have no active heating or cooling systems.[42] There are no radiators, no air conditioners, there is no need even for a wood-burning stove. The heat they require is produced by sunlight coming through the windows and by the bodies of the people who live there.

This sounds like a formula for misery, but a study of over 100 passive homes showed they had a mean indoor temperature of 21.4° during the cold German winter:[43] higher, in other words, than the average temperature of houses in the United Kingdom. Even the unoccupied homes in the study stayed at 17°. In summer, the temperatures seldom rose above 25°.[44]

There is nothing magical about these constructions, and they rely on little in the way of innovative technology. The builders need only ensure that the 'envelope' of the house – the bit that keeps the weather out – is as airtight as possible, and contains no 'thermal bridges'. A thermal bridge is material that conducts heat easily from the inside of the house to the outside. At every point – even where the walls meet the ground or the roof – contact with outside temperatures must be interrupted by insulating materials.

This doesn't mean that the house should become a sealed box. Our slow but steady progress in stopping up leaks is one of the suspected causes of the rising incidence of asthma. Passivhauses have automatic ventilation systems which ensure that all the air in the house is changed once every three or four hours.[45] They use a heat exchange system: the cold air entering the house is passed over the warm air leaving it, capturing about 80 per cent of its heat. (This uses energy, but far less than a conventional heating system.) To make this system more effective still, you can draw the fresh air through pipes in the soil,[46] which in the winter remains warmer than the air, and in the summer stays cooler. The important feature is that the heat exchange systems are the *only* points through which air passes.

The windows are also important. In the northern hemisphere, they should be mostly south-facing and carefully matched to the size of the house: if they are too small, the house becomes too cold, too big and it gets too warm. They should provide about one third of its heating.[47] To absorb more heat than they release, they need to be triple-glazed and super-insulated. The houses should also have a high 'thermal mass': the materials should be able to store heat, so that warmth absorbed from the sun during the day can continue to radiate at night.

Altogether, when fitted with efficient appliances, passivhauses should save around three quarters of the energy of an ordinary modern home of the same size.[48] Remarkably, they are not much more expensive. The extra building costs should amount to 10 per cent of the total or less.[49] A development of twenty homes in Freiburg, with a measured energy saving of 79 per cent, for example, cost just 7 per cent more than a typical building of the same type.[50] Some designers say they have brought the extra costs down to zero.[51] The reason why the costs remain so low, even though the building standards are higher and some of the materials (like the windows and the insulation) are more expensive, is that you don't have to install a heating or cooling system.

If every house were magically transformed into a passivhaus by 2030 we would come close to reaching our 90 per cent target for the housing sector, even before we changed the sources of the remaining energy they use. Of course, unless we knocked down and rebuilt our entire housing stock between now and then (which would itself produce a tremendous amount of carbon), this cannot happen. But it is shocking to see how slow the uptake has been even for new homes.

There are now around 4,000 passive houses in Germany, 1,000 in Austria[52] and a few hundred elsewhere. In the United Kingdom, with a couple of possible exceptions, they are confined to a development in south London by the man I attacked in the introduction, Bill Dunster. Happily, his architecture is more reliable than his claims about wind turbines. The Beddington Zero Energy Development (BedZed) he designed does have a heating system, but it uses just 10 per cent of the energy that ordinary buildings of the same size would need,[53] and this is supplied by woodchips from trees pruned by the council.[54] (When the chips are burnt they also generate some of BedZed's electricity.)

Dunster has incorporated some clever features. The lagging for the hot water tanks, for example, is not wrapped around the tanks themselves, but lines the cupboards in which they sit. People can use the cupboards to air their clothes, while losing very little of the heat from the tanks. If they have been away and want to heat the house quickly, they can leave the vents in the cupboard doors open for an hour or two.[55] He has also sought to save water (taking some of it from the roofs) and encourages people to use their cars as little as possible. The development incorporates offices (also built to the passivhaus standard), a car pool and walking and cycling schemes. Though every flat has its own garden, BedZed (which contains 99 homes and workspace for about 100 people) is as densely packed as developments in central London. Houses like this could be put up almost anywhere.

Like that of many other rich nations, the British government wants to see a massive number of new homes built – 1.2 million by 2016 – to accommodate the people desperate to escape from their families. It is hard to see why all these new homes cannot be built to passivhaus standards. Indeed if they are not, the housing sector's contribution to climate change is likely to rise greatly by 2030[56] as – all else being equal – more homes means more energy. But the only thing that could encourage a widespread construction of low-energy homes is a set of building regulations that insist it must happen. This means setting a date for universal passivhaus standards – I suggest 2012 – and ratcheting up the building codes every year between now and then. This would provide a massive incentive for the construction industry to invest in some research and development and train its workers properly. At the moment, our builders get away with practices very similar to those which prevailed in 1900, when my own house was built.

If our governments refuse to attend to the building codes, they will vitiate not only any rationing scheme we might hope for, but also any meaningful cap on carbon emissions. When people live in homes which are incapable of sustaining a low-carbon life, they will not tolerate an attempt to cut their emissions by 30 per cent, let alone by 90 per cent. A carbon cap without proper building regulations really does amount to a demand that people huddle around a smouldering log to keep warm.

But the problem of new housing is an easy one to solve – by comparison to the question of what we do with existing homes.

Few of the new homes being built in this country replace houses that exist already. While around 160,000 are going up in the UK every year, only 15,000 – 0.06 per cent of the total stock of 25.5 million – are knocked down.[57] This means that it would take almost 1,700 years to replace the houses standing today. In 2005, Oxford University's Environmental Change Institute proposed that the rate be increased fourfold if there were to be any hope of meeting the more modest target (a 60 per cent cut in carbon emissions by 2050) set by the government.[58] This attracted great controversy, not least among people who alleged that the carbon costs of demolishing and replacing these buildings would outweigh the carbon saved by more efficient homes. But something has to be done, because some 2 million of the houses here appear to be beyond cost-effective repair.*[59]

There's scarcely a clearer sign of the low priority given to energy-efficient housing than the complete absence of research into the energy costs of knocking down inefficient homes versus the energy costs of leaving them standing. The Environmental Change Institute's report cites two papers to support its contention that

---

* They have an SAP rating of under 30, out of a possible 120.

When an old, inefficient building is replaced with a new, efficient one, the ... energy in the construction process will be offset within a few years.[60]

But neither paper says any such thing.*[61,62]

Either way, we have to assume that the great majority of the houses standing today will still be standing in 2030, so most of the work, with all the difficulties this entails, will have to be done within existing homes. I've already suggested the means by which this might be achieved: building regulations governing their refurbishment. The rules would specify, for example, that if ever a floor were lifted, a wall rendered or a roof replaced, the restoration would have to be accompanied by insulation, the plugging of leaks and the sealing off of thermal bridges. While refurbishment will never bring our existing homes up to passivhaus standards, the government estimates its 'technical potential' for saving energy as between 40 and 42 per cent across the whole housing stock.[63] This was described by the House of Lords, when it investigated these questions, as 'relatively conservative'.[64]

Of course, an unintended consequence of much stricter rules governing the refurbishment of houses is that it could discourage any refurbishment at all, thereby ensuring that they remained even less energy efficient than they would otherwise have been. A carbon rationing system will provide a powerful incentive for refitting, but because of the expense it might be necessary to supplement it with some other measures. For example, governments could offer a rebate on stamp duty – which is the tax people must pay when they sell their homes – to cover part of the cost of restoration.[65]

In the UK, the companies which supply gas and electricity are obliged to spend some of the money they make on helping householders reduce their energy bills.†[66] While it's not clear whether this results in significant savings,‡ it's quite a good example of an incentive

---

* One of them provides a direct comparison only for 'wooden materials'; the other investigates only the energy costs of insulation.

† This is called the Energy Efficiency Commitment.

‡ In theory it reduces carbon emissions by 0.7 million tonnes – out of the 40 million our homes produce[67] This is regrettably no more than half the extra carbon emitted by the new buildings going up every year. But even this saving could be illusory: the government has no means of telling how much of it is lost through the rebound effect.[68]

which doesn't cost the government anything. But, by comparison to programmes in other countries, our schemes are hopeless. Soon after Angela Merkel became Chancellor of Germany in November 2005, she announced that her government would be spending the equivalent of £1 billion a year to ensure that 5 per cent of the homes built before 1978 were refurbished to meet high energy efficiency standards: within twenty years every house in the country will be airtight and well-insulated.

A particular problem is the troubled relationship between landlords – especially private landlords – and their tenants. Because, on the whole, the tenant pays the electricity and heating bill, the landlord has no incentive to make the house and its appliances more efficient. For this reason, a rationing system could be grossly unfair on tenants: they might have a powerful incentive to improve the performance of their homes, but no means of doing so.

Various incentives have been proposed for encouraging landlords to improve their buildings, and the government has even introduced a tax break for them (they can set the costs of insulation against income tax[69]). But I don't think we need waste too much sympathy. Private landlords are already obliged to install various safety features (such as firedoors and fire escapes) before they rent out a home, and everyone recognizes that they should carry these costs. It seems to me that any home to be let should first pass a series of energy tests, and the landlord should be obliged to pay for the improvements.

However much we enhance the fabric of our homes, there's a danger that the energy consumption inside them could keep rising. This is because of the explosive growth of electronic gadgets.

In the thirty-one years to 2005, the use of electricity for lights and appliances in our homes rose by 2 per cent a year.[70] Houses now consume a quarter of the UK's electricity. This is partly because we own more of every kind of technology (between 1990 and 2003, for example, the number of households owning video recorders rose from 59 per cent to 88 per cent [71]), partly because our televisions, fridges and washing machines are getting bigger, and partly because there are more kinds of technology to own. While the efficiency of some gadgets – such as fridges and freezers – has greatly improved, others have become worse. A large plasma TV, for example, uses almost five times as much electricity as an ordinary model with a cathode ray tube.[72] Until recently, telephones received all the electricity they needed from the few milliamps the phone companies send down the line. Now it is hard to buy one without both a battery and a plug.

Daftest of all is the electricity we use for no reason whatever: when our appliances are switched off. According to the British government, around one million tonnes of carbon emissions a year are caused by equipment in homes and offices left in 'standby' mode: plugged into the wall but not operating.[73] This uses about 2 per cent of all our electricity.*[74] And the problem could become worse, as the digital decoders – or set-top boxes – for our televisions, which are rapidly becoming universal, are designed never to be unplugged.[75]

The only sector in which energy consumption has fallen substantially (by 15 per cent between 1990 and 2003[76]) is cooking, and this is a false saving. Our meals require less energy to prepare only because they have already been prepared by someone else.

The table below gives a breakdown of our use of electricity in the home.†

In every one of these sectors, the waste of energy is scarcely believable. Compact fluorescent lightbulbs, which have been commercially

gadget	electricity consumption (terawatt hours per year)
consumer electronics (TVs, computers, phones, etc.)	10.4
washing machines, dryers and dishwashers	11.8
cookers, kettles and microwaves	11.9
lights	17.4
fridges and freezers	17.5
Total	73.0

Source: Environmental Change Institute, Oxford University.[77]

---

* In 2004 power stations in the UK emitted 47 million tonnes of carbon.
† The figures are for the nation as a whole.

available for over twenty years, burn around a quarter of the energy of ordinary incandescent bulbs, yet so far we use them at an average rate of just 0.9 per household.[78] It is still almost impossible to buy LED (light-emitting diode) bulbs, though these are even more efficient than compact fluorescents. A fridge or freezer which uses vacuum-insulated panels to stay cold burns about 12 per cent of the energy of the average model used today,[79] but simply cannot be found for sale through the usual channels in the United Kingdom. Because electricity remains so cheap, and the incentives to conserve it are so slight, the great potential of new technology is mostly being squandered.

A rationing system would provide a permanent incentive to seek out better equipment, and the manufacturers, even if they were subject to no regulatory restraints, would try to supply it. But people can make sensible choices only if they know exactly what they are buying. While energy labelling in the European Union has improved, the manufacturing companies have done their best to make it as confusing as possible.

For example, companies selling fridges and freezers in the EU are obliged to give them labels showing how much electricity they consume. At first the sequence ran from A to G, with A being the most efficient, and the expectation was that the standard for every category would be steadily raised as time went by. But instead of doing this, the European Commission, under 'political pressure by the manufacturers'[80] simply added two new categories to the scheme: A+ and A++. So the appliances now sold as grade 'A' should in fact be grade 'C', and the true A (A++) is nowhere to be found. Energy-conscious people, most of whom remain ignorant of this duplicity, buy the fake 'A' equipment, believing it sits at the top of the range. There are no official energy labels at all for products like TVs and computers.

But even the weak guidelines we do possess could be at risk. In October 2005, a group of manufacturing countries, including the United States, China and South Korea, sought to persuade the World Trade Organisation that all energy labels are a 'barrier to free trade' and should be made illegal.[81] The negotiations are continuing.

As well as labels, Europe does have some mandatory minimum

standards, for goods such as fridges and freezers. Their utter feebleness is demonstrated by the fact that – despite the dire predictions of the manufacturers – the price of the regulated gadgets has continued to fall.[82] In other words, the companies have been able to meet the new standards at very little cost, and the market could have borne a much tougher requirement.

In Japan and Australia by contrast, the government finds the most efficient model and insists that by a certain date all others must match it. The House of Lords alleges that because of the timidity of our rules, Europe is now becoming 'a dumping-ground for less efficient goods'.[83]

It is also hard to understand why we cannot be allowed to see how much electricity our appliances are using. It would cost manufacturers next to nothing to install a panel on their gadgets – a bit like the digital thermometer in a fridge – showing how much electricity it is consuming. A study of households whose cookers were fitted with electricity meters found that they reduced the energy they used for cooking by an average of 15 per cent.[84]

Two studies show that a 'smart meter' measuring the electricity used by the whole house reduces consumption by around 12 per cent.[85] A smart meter is a small panel put in a place where it can be easily seen – perhaps just inside the front door – with a clear digital display, preferably in a useful unit such as pence per hour. Some varieties can tell you how much electricity each appliance is using. At the moment, meters exist entirely for the benefit of the supplier. They are generally inaccessible and, when you have fought your way past the junk they are buried under, almost incomprehensible. One study shows that more than 50 per cent of adults do not know where their gas and electricity meters are, and 45 per cent can't read them once they have found them.[86]

Just as shops would be able to make more money if they never put prices on their products, it is in the interests of the electricity suppliers that we should have no idea how much we are consuming. In Ontario, Canada, the government has ruled that smart meters will be installed in every home by the end of 2010.[87] They will cost around 250 Canadian dollars – £100 – each. In the UK, by contrast, the government – perhaps as a result of lobbying by the electricity companies –

obstructed an attempt by the European Union to start introducing them.[88] The energy review it published in July 2006, however, suggests that this policy has changed.

A company called More Associates has taken the idea a step further. A meter just inside the front door not only displays the amount of electricity your house is using, but also contains an off switch: as you go out, you can turn off the entire house, except for the gadgets you have already selected to stay on permanently.[89]

And why should we not also be allowed to see how much carbon dioxide we have used? Food manufacturers now tell us how many calories their products contain. It would surely be no more difficult for the suppliers of gas and electricity to print our carbon dioxide consumption on the bills they send us.

So how far and how fast can we cut the carbon-based energy our houses use? The Environmental Change Institute set out to see whether the British government's target of a 60 per cent reduction by 2050 could be met. It found that even with better heating and a slight growth in the number of gadgets, it could be – but only just.

In reality, these targets are approaching the extreme end of the policy envelope: it would be close to inconceivable to plan for tougher standards ... on this timescale.[90]

And this was *after* the Institute had assumed that almost every house would have two 'low or zero carbon technologies' fitted to it, which means solar panels, solar hot water, heat pumps, wood burners, wind turbines or combined heat and power units.*

It seems to me that this might be slightly too gloomy. The report assumes, for example, that nothing can be done about the growth in the number of households. But as the climate-change campaigner George Marshall has pointed out, a programme that helped single elderly people to leave their big, draughty homes and move into smaller, warmer flats could both reduce the pressure for new building and cut the number of deaths in winter.[91] Even so, the constraints the Institute identifies have to be taken seriously. There are physical limits to the degree to which the old housing stock can be improved. Thanks

---

* I will explain all this in Chapter 6.

to yet another failure on the part of government – in this case to train skilled fitters[92] – even when the right incentives exist, there aren't enough people to do the work. However well the new means of saving energy are explained, some people just won't follow them.

In other words, there are limits to energy efficiency, even within a cap and rationing system. The Environmental Change Institute's figures suggest that the maximum reduction of energy use in housing by 2030 is likely to be a little over 30 per cent: around one third of my target.

What this means is that most of the cut will have to be made by changing the sources of the energy our buildings use: in other words by producing fuel and electricity whose carbon content is as low as possible. This task, which is much harder than many people – especially the advocates of renewable energy – have led us to believe, is the subject of the next three chapters.

5

Keeping the Lights On

There wave on wave imbued with power has heaved,
But to withdraw – and nothing is achieved;
Which drives me near to desperate distress!
Such elemental might unharnessed, purposeless!

Faust, Part II, Act IV[1]

As with oxygen, we notice electricity only when it fails. Vaguely aware that most of what we do would be impossible without it, we seldom have to wonder where it comes from, how it works or what we would do if it were no longer available. Yet its steady supply would astonish Mephistopheles.

Unlike almost all the other commodities we buy, which can be stockpiled and then delivered when we want them, electricity must be produced at the very moment of demand. This is because it is difficult and expensive to store.[2] And if either too much or too little is produced, the voltage and frequency fluctuations will crash the country's computers. If supply falls below a certain level, the whole grid collapses. So not only must it be made when we want it; it must also be made in precisely the quantities we demand.

This would not be too difficult if our demand remained constant. But it oscillates wildly. On a summer's night in the United Kingdom, for example, we need less than 20GW of power-generating capacity (a GW is a gigawatt, or one billion watts). On cold winter evenings, when about half the population returns from work at around the same time, and turns on the lights, the kettle, the television, the power shower and the electric heaters, over 60GW has to be found.

Exceptional events, which cause us to synchronize our behaviour more precisely, can push this even higher. During the 1990 World Cup semi-final, for example, demand rose by 1.6GW within a couple of minutes at half-time and at full-time, and by 2.8GW after the penalty shoot-out.[3]

Why? Because as soon as there was a break, most of the population got up to put the kettle on. 2.8GW is more than twice the 'load' of the biggest power station in the UK.[4] Because our televisions and lights were burning already, the system was straining even before these extra demands were made. Someone had to ensure that the extra power was found at the very moment it was required, and not just *some* extra power, but a quantity matched exactly to the demand.

What this means is that the electricity companies must study the behaviour of their customers, chart their historical use of electricity, anticipate public holidays and national events, read the weather forecasts and the television schedules and study the ratings, and watch the penalty shoot-outs in order to determine when the entire country will lever its collective backside off the sofa.

To respond to these fluctuations in demand, they need constantly to be bringing power plants in and out of production. The 'baseload' – the 20GW we use all the time – is supplied by nuclear reactors and large gas-powered plants. As more electricity is required, coal-burning power stations and smaller generators are brought online. Some plants will be kept out of production all through the summer, and fired up as demand rises in the winter. Some are held in 'spinning reserve': operating, inefficiently, at just part of their load, but ready to be brought up to full power in seconds. A cable between France and the United Kingdom allows us to import 2GW of electricity when we fall short. Some factories strike special deals with the electricity companies: in return for a rebate, they agree to cut their demand when the system comes under strain.[5]

Most dramatically, the UK has three 'pumped storage' plants, which provide our only cost-effective means of storing power. They each consist of two reservoirs, one at the top of a mountain; one close to the bottom. When electricity is cheap, which means when demand is low, it is used to pump water from the bottom reservoir to the top one. When there is a requirement for a sudden surge in production, the gates of the top reservoir are opened, and the water pours through turbines back down to the bottom. The pumped storage plant at Dinorwig, in north Wales, can produce 1.7GW of power for five hours.[6] It responds within fifteen seconds.[7] I like to picture a man in a booth with his television on. As the match draws to an end, the phone rings. 'It's the last penalty. Open the gates.' He pulls down a great red lever, and the water roars out of the upper reservoir just as the ball thumps into the corner of the net. I'm sure that in reality it's all done automatically.

Seeing how this system works, I am struck by the thought that perhaps we shouldn't be surprised to learn that electricity companies have been reluctant to invest in uncertain power sources such as wind and waves. The system is already so finely balanced that the instinct of anyone who runs it must be to minimize further complexity. But something has to be done, because this miracle is sustained only by burning vast amounts of fossil fuel.

In the United Kingdom, our electricity comes from the sources show in the table below. Oxford University's Environmental Change Institute estimates that while the carbon content of the *energy* we use in the home can be cut overall by 60 per cent by 2050, our *electricity* consumption can be reduced by just 16 per cent in the same period.[8] This is depressing, especially when you realize that the institute's timetable is twenty years longer than mine. But it assumes that nothing is done to suppress the growing demand for new and often bigger

fuel type	percentage of electricity generation
gas	41
coal	33
nuclear	19
renewables	3
imports	2
oil	1
other fuels	1

Source: UK Department of Trade and Industry.[9]

electrical appliances. My belief – and I hope that I am not being unduly optimistic – is that a rationing system accompanied by either regulation or an effective public information campaign (telling people, for example, that large plasma TVs consume five times as much electricity as other models) would go some way towards reversing the current trend. I am going to assume that it's possible to reduce the demand for electricity by some 25 per cent by 2030. This means that I must still find a way of cutting its carbon content by more than 80 per cent.

Coal contains an average of 24.1 kilograms of carbon per gigajoule of energy, while natural gas contains just 14.6 kilograms.[10] So, if all else were equal, burning gas rather than coal would produce about 40 per cent less carbon dioxide for every watt of electricity it generates. But coal is even worse than this suggests. A modern gas-burning power station turns about 52 per cent of the energy its fuel contains into electricity.[11] The best coal-fired generators have an efficiency of just 40 per cent.[12] Partly because we still use coal, and partly because some of our gas turbines are of rather antiquated designs, the average efficiency of power stations in the United Kingdom in 2004 was just 38.5 per cent.[13]

Plainly, one significant step towards a lower-carbon economy is to swap the two fuels. This, though not for environmental reasons, has already happened to a large extent in the UK. In 1970, we consumed 176 million short tons of coal (a short ton is 2000lbs), and in 2003 just 69 million.[14] The switch to gas is one of the major reasons for the reduction in the United Kingdom's carbon emissions: without it we would have had little hope of meeting our obligation, under the Kyoto Protocol, to produce 12.5 per cent less carbon in 2012 than in 1990.

Unfortunately, after a couple of decades of substitution in the rich nations, it looks as if we are starting to travel in the opposite direction. In 2025, according to the US government's Energy Information Administration, the United States will burn 40 per cent more coal than it does today.[15] China intends to treble the electricity it produces from coal by 2020.[16] The British government expects – if the market has its way – a major expansion of coal burning in the UK after 2020.[17]

It is not hard to see why the world is returning to the Coal Age. Natural gas supplies in North America have already peaked and are going into decline.[18] In Europe, wholesale gas prices tripled between 2003 and 2006.[19] This is partly because North Sea gas has also begun to diminish; and partly because the government of Russia temporarily restricted supplies to Eastern European countries,[20] while the gas companies that control the pipelines have been limiting the supply to Western Europe.*[21]

Some people are concerned that global gas supplies could soon follow North America's over the brink and into decline. The UK's Parliamentary Office of Science and Technology, for example, citing an organization called the Association for the Study of Peak Oil and Gas, predicts that 'the global gas production peak will occur by 2020–2030.'[22] I find this unlikely.

The Royal Commission on Environmental Pollution expects that 'global production will not peak until 2090',[23] though the expert it cites, Professor Peter Odell, has made predictions about the oil price which have proved to be optimistic.[24] Shell proposes that known gas reserves will be much bigger in 2020 than they are today,[25] but I have learnt to be wary of the oil companies' projections. Perhaps a more reliable prediction is the one made by the Geological Society of London. Within the next fifty years, it suggests,

oil will probably experience a supply peak . . . while natural gas will not . . . although projection of historic growth rates suggest production constraints may arise around mid-century.[26]

Coal, by contrast, faces no such restrictions. The International Energy Agency says the world currently has around 1 trillion tonnes of recoverable coal – enough for 200 years of production at present levels.[27] The Geological Society points out that a price increase of only $10/tonne would, in effect, double the world's economic coal reserves.[28] The Norwegian oil company Statoil estimates that under

---

* The energy regulator Ofgen, for example, alleges that the gas companies kept the flow through the 'Interconnector' – the pipe across the North Sea – to just 60 per cent of capacity in the winter, when we need gas most. At the beginning of 2006, Russia cut its gas supplies to Ukraine, which in turn supplies much of the rest of Europe, in order to force a change in the contract.

the Norwegian seabed alone there are 3 trillion tonnes of coal, though at the moment there is no viable means of extracting it.[29] Ninety per cent of the remaining energy reserves in the US are coal.[30]

There are means of increasing the security – and therefore the economic longevity – of our gas supplies. The most effective is an increase in the amount of storage. In the United Kingdom, for example, we store only fourteen days' supply, by contrast to a European average of fifty-two days.[31] This means that our gas supplies are less reliable than other people's, which increases the pressure – from bodies such as the Confederation of British Industry – to switch to other kinds of fuel. Yet we have plenty of potential reservoirs, in the form of salt beds and old, depleted gas fields under the North Sea. You can store gas simply by pumping it back into the rocks. This is already done on a small scale in the Rough gas field in the southern North Sea,[32] and on a larger scale in Ohio, West Virginia, Pennsylvania and New York State.[33,34]

I am not suggesting that burning gas will save the biosphere; simply that a return from gas to coal will greatly accelerate its destruction.

But even if we continued to produce most of our electricity from burning fossil fuels, we could, in theory at least, cut carbon emissions by 80[35] or 85 per cent.[36] The technology which would make this possible is called 'carbon capture and storage'.

This means stripping the carbon dioxide out of the fuel either before or after it is burnt, and burying it in the hope that it will stay where it's put. Like so much to do with the electricity industry, it sounds ridiculous. But in some places it is already happening.

Already, Statoil scrubs 1 million tonnes of carbon dioxide a year from the gas it extracts from the Sleipner field in the North Sea, and pumps it into a saline aquifer (a pocket of salt water) under the seabed.[37] BP is doing something similar in Algeria.[38] (The carbon dioxide is an impurity which reduces the value of the natural gas.) Since 1954, a Canadian company called EnCana has been using carbon dioxide to flush the remaining oil out of a depleted reservoir in Saskatchewan. Though at first EnCana had little interest in what happened to the gas, about three quarters of it has stayed underground.[39]

There are several means of extracting carbon dioxide from the exhaust gases of a power station. The most likely technology has been used for other purposes for over sixty years.[40] It's called 'amine scrubbing'. The gases are bubbled through chemicals called ethanolamines, which absorb between 82 and 99 per cent of the carbon dioxide.[41] The chemical mixture is then heated to release the carbon,[42] which is piped away for burial, so that the amines can be re-used.

After the gas has been captured, it is compressed and then pumped down a pipeline and into an underground store. Several kinds of geological feature seem capable of holding it almost indefinitely. Below is a list of the world's possible reservoirs and the amount of carbon dioxide they could hold.

reservoir	global capacity (billion tonnes)
old oil fields	125
unmineable coal seams	148
old gas fields	800
saline aquifers	400–10,000

Source: UK Department of Trade and Industry.[43]

The wide range of the figures for saline aquifers is rather worrying. The world's power stations produce around 10.5 billion tonnes of carbon dioxide a year.[44] In principle, if the higher estimates are correct, you could store all the carbon our electricity generators will produce in the next twenty-four years several times over. But, as I will show in a moment, even doing it once is difficult.

There are good reasons to suppose that once carbon dioxide has been properly buried in the right sites, it will stay where it is put. Most of the natural gas reservoirs we exploit today have remained stable for millions of years. The IPCC says that after 1,000 years, less than 1 per cent of the carbon dioxide buried in appropriate reservoirs is likely to have escaped.[45]

A fortunate property of carbon dioxide helps to keep it where it is: at 800 metres or more below the earth's surface, the pressure turns it 'super-critical': it behaves more like a liquid than a gas.[46] According to the International Energy Agency, about 80 per cent of the world's oil fields, into which some of the gas could be pumped, occur at depths of over 800 metres.[47] Capturing and burying carbon dioxide costs anywhere between £12 and £160 a tonne, depending on whose estimate you believe.[48] Capturing carbon from coal-burning power stations costs more than capturing it from gas generators, as coal contains more carbon per tonne. Altogether, power stations in the United Kingdom produce 172 million tonnes of carbon dioxide a year.[55] A table in Chapter 7 compares the costs of saving carbon by all the different means I discuss.

When I first heard about this technology, I reacted with hostility. My first thought was that it couldn't possibly work: the gas would surely leak from the aquifers. This fear has now been laid to rest. Then it struck me that the energy (and therefore the carbon) costs of extracting, compressing, transporting and burying it would outweigh any savings it permitted. But the IPCC maintains that while a power plant would need to burn between 10 and 40 per cent more fuel to cover these energy costs, the net carbon saving would still be 80–90 per cent:[56] the carbon dioxide from the extra gas or coal it burnt could also be buried.

Then I expressed the concern that the promise of carbon capture and storage could provide an excuse for the fossil fuel companies to pursue their business as usual. Surely a much safer means of dealing with fossil fuels is to leave them in the ground? But this argument was countered in an interesting fashion by Jonathan Gibbins at Imperial College, London. The rules required to keep fossil fuels unmined, he maintained

probably have a much lower reliability and . . . longevity than geological storage. Storage also requires a one-off effort up front, not sustained dominance of global policy.[57]

In other words, once you have buried your carbon dioxide, you can more or less forget about it; but if valuable assets remain under the ground, you will need constant enforcement to prevent them from being extracted. Carbon capture and storage is more politically stable than economic restraint.

But there are three arguments against it which do bear some weight.

The first one is that almost all the cost estimates I've read are accompanied by a second and lower set of figures: for a process called 'enhanced oil recovery'. The basic mechanism is the same – you inject carbon dioxide into old oil fields – but in this case you do it partly in order to squeeze the dregs out of them (as EnCana has done in Saskatchewan). The gas dissolves in the oil, reducing the oil's viscosity. As it is forced through the reservoir, it drives the oil into the production wells.[58] The money this makes explains why enhanced oil recovery costs less than straightforward carbon burial. Less of the carbon dioxide stays underground, as some of it will escape with the oil. And more oil than was otherwise available comes to the surface. A report for the US Department of Energy suggests that carbon dioxide injection could effectively quadruple the country's oil reserves.[59] As most oil is used in vehicles, and there is no viable means by which the carbon they produce can be captured and buried, there's a danger that using carbon dioxide for enhanced oil recovery could *increase* its concentrations in the atmosphere. If burying carbon is to be used as a means of tackling climate change, it cannot also be used as a means of recovering oil.

The second argument is that capture and storage helps to revitalize the coal industry. Already, on the strength of nothing but speculation about what might one day be possible, the industry's boosters have managed to drum the words 'clean coal' into our ears. While the carbon dioxide from coal-burning power stations might one day be buried, in every other respect coal is likely to remain one of the world's most destructive industries. If you doubt me, try standing on the edge of an open-cast mine and saying the words 'clean coal'.

In the Appalachian Mountains in the eastern United States, the coal companies, always innovative when it comes to planetary destruction, are now using a method of extraction they call 'mountaintop removal'. They simply blast the tops off the mountains and bulldoze the rubble into the valleys, turning a fissured sierra back into a plateau. Already they have buried 1,200 miles of streams.[60] When I tell my friends that if I were forced to choose between nuclear power and coal, I would pick nuclear, they go berserk. I invite them to take a look at some pictures of the mines in West Virginia.[61]

But, at the risk of playing the same speculative game as the coal barons, I should point out that there is a technology which could one day realize the promise of clean coal. This is called 'underground coal gasification'. Holes are drilled into a coal seam and air and steam are pumped into it. The coal is 'gasified', releasing methane and hydrogen, both of which can be burnt in power stations. Carbon can be extracted from the gases either before or after they are burnt. The technique requires no major excavations, no tailings or slag heaps, and no children breathing dust in narrow galleries.

It has been practised since the 1950s in Uzbekistan and tested, with some success, in Australia, Spain, China, the United States and the United Kingdom. It's now possible to gasify seams as far as 600 metres below ground.[62] The deeper it is done, the better, as it is then less likely to contaminate freshwater aquifers.[63] Like all coal burning, it releases other pollutants, such as sulphur and nitrogen oxides and heavy metals, so these too need to be stripped out of the gases which emerge from the boreholes. The costs are comparable to burning coal in a modern power station.[64]

But it is critical that underground coal gasification is used *only* with carbon capture and storage. Otherwise, because it can be applied to seams too narrow to dig, it simply opens up even more coal than was otherwise available. In the United Kingdom, for example, eleven times as much coal can be exploited by underground gasification as by mining or quarrying.*

The third viable argument against carbon capture and storage is that, like almost anything we choose to do, it prevents other options from being pursued. In this case, it means that the waste heat from electricity production will not be available for warming our homes. This is because capture and storage makes economic sense only in very large power stations, while 'combined heat and power' (which I will explain in Chapter 7) makes economic sense only in smaller ones.[67]

---

* The US Energy Information Administration estimates that the UK has 1.65 billion short tons (1.5 billion tonnes) of recoverable coal reserves.[65] The British government says that 'UK coal resources suitable for deep-seam UCG on land are estimated at 17 billion tonnes (300 years' supply at current consumption) and this excludes at least a similar tonnage where the coal is unverifiable for UCG.' There are also vast resources under the sea bed.[66]

But even if we decide that catching and burying carbon dioxide from fossil fuel burning is the best way of decarbonizing our electricity supply, it cannot be used everywhere. A power station has a life of about forty years.[68] Unless it has been built with carbon capture in mind, the necessary equipment is difficult to bolt on. The plant should be built within 500 kilometres of the place where the carbon will be buried, because the transport costs increase with distance. Enough space needs to be left to fit the extra pipes and valves and build the capture plant.[69] It simply won't be possible in many of the world's power stations, including many of those being constructed now. Partly because of this, but partly because of the time taken to design, develop and test the new technology, the International Energy Agency (IEA) believes that

Large-scale carbon capture and storage is probably ten years off, with real potential as an emission mitigation tool from 2030 in developed countries.[70]

At first sight, in other words, it appears to be too far away to make a major contribution to meeting our target. But as I will show at the end of this chapter, estimates like the IEA's could be unduly pessimistic. I have come to believe that this technology, alongside others which have been judged 'too far away', can, with sufficient political commitment, be widely deployed long before 2030. The difficulties I have encountered while investigating the other technologies have persuaded me that carbon capture and storage – while it cannot provide the whole answer – can be and must be one of the means we use to make low-carbon electricity.

Here begins the section of the book I have been dreading most: a discussion of nuclear power. I hate this topic partly because it is charged with more anger than any other; partly because every fact is fiercely contested. However much reading you do, you still don't know what or whom to believe.

The particular problem environmentalists face is that the movement itself arose partly as a result of concerns about nuclear energy, which were closely linked to fears about nuclear weapons proliferation. Anything which suggests that you are giving serious consideration to nuclear power risks being perceived as an attack on environmentalism. Indeed, so large does this issue loom among the greens that climate change is often subordinated to it. Several organizations have published reports showing that there is no need to build new nuclear power stations as the old ones become redundant, because renewable energy can fill the gap.[71,72,73] The danger is that we end up replacing nuclear power rather than replacing fossil fuels.

The link between nuclear electricity and nuclear weapons is a real one. There is a grim symmetry in the technology's development. In the first nuclear nations, nuclear power generation was a by-product of nuclear weapons development. In the later nuclear nations, nuclear weapons development was a by-product of nuclear power. Every state which has sought to develop a nuclear weapons programme over the past thirty years – Israel, South Africa, India, Pakistan, North Korea, Iraq and Iran – has done so by diverting resources from its civil nuclear reactors.[74,75] The more nuclear material the world contains, the more weapons it is likely to develop, and the more widespread they will become.

When considering all the other technologies I will discuss in this book, you can judge them by three criteria: environmental impact, feasibility and cost. But in this case we have to consider another factor. How do you rate the threat of climate change against the threat of nuclear war? One is certain to happen – indeed is happening already. The other is just as devastating – perhaps even more so – but less certain. Any one contribution to the world's stockpile of nuclear materials might not make any difference to the possibility of war, though the total increase in volume appears to make it more likely.

One fact which does seem pretty certain is that every nuclear power station leaks radiation into the environment. As well as their routine emissions into the air and the sea, the nuclear generators are surrounded by dumping scandals. In the United Kingdom, for example, hardly a year goes by without some new and terrible revelation about the nuclear complex at Sellafield in Cumbria. In 2004, the European Commission took the British government to court over Sellafield's refusal to let its inspectors examine one of its dumps.[76] (You may remember that we went to war with Iraq over something like this.) It's hardly surprising that the complex wanted to keep them out: in 2003 they discovered a pond containing 1.3 tonnes of plutonium, which had been sitting there, unacknowledged and unchecked, for thirty years.[77] In 2005, investigators found that a pipe at the complex had been leaking, undetected, for over eight months, spilling nitric acid containing some 20 tonnes of uranium and 160 kilograms of plutonium.[78]

In 1997, the operators of the power plant at Dounreay on the north coast of Scotland admitted that for many years they had been dumping its waste into an open hole they had dug above the crumbling coastal cliffs. The shaft had already exploded once – in 1977 – scattering plutonium over the beaches, but the UK Atomic Energy Authority, which runs the plant, hadn't bothered to tell anyone.[79] The authority promised that there would be no more cover-ups. But less than a year later it was forced to admit that it had dug a second hole in the cliffs, into which it was still dumping unsealed nuclear waste.[80]

There are two reasons why cheating is so common in the nuclear industry. The first is that it is much cheaper to handle radioactive materials badly than to handle them well. The second is that the nuclear operators have the perfect excuse – security – for withholding inconvenient facts from the public.

The release of radioactive materials, among them the most toxic element on earth (plutonium), into the environment is, of course, dangerous for human beings. The number of deaths it causes is as controversial as every other set of facts about nuclear power. But it seems likely that, as a result of both routine and accidental discharges, some people die of cancer in most nuclear nations every year. A meltdown or a successful attack by terrorists, though improbable in the rich nations, would kill more: the estimates for grave illness caused by radiation from the Chernobyl disaster in 1986 range all the way from a few thousand to several million. It is probable that several thousand people will die prematurely as a result of the accident.[81] But the grim moral accountancy which must inform all the decisions we make obliges me to state that nuclear power is likely so far to have killed a much smaller number of people than climate change.

Cheating, because it is so much cheaper, also governs the intentional disposal of nuclear waste. In theory, it can be buried safely. I found the technical report produced by the Finnish nuclear authority,

Posiva, convincing.[82] The spent fuel is set in cast iron, which is then encased in copper and buried at the bottom of a borehole. The hole is filled with saturated bentonite, which is a kind of clay. Posiva's metallurgists suggest that under these conditions the copper barrier would last for at least a million years.[83]

The danger is that Posiva's good example is used as a Potemkin village by the rest of the industry: a showcase project which creates the impression that the problem has been solved but behind which all the usual abuses continue. The government of the United Kingdom still has no plans for the long-term disposal of its nuclear waste. This seems to breach the most fundamental environmental principle, one that children are taught as soon as they are old enough to understand it: you don't make a new mess until you have cleared up the old one. One of the reasons for this omission is that the government body responsible for finding a place to bury the waste squandered all public confidence by choosing a site (Sellafield) for political rather than geological reasons.*[84] In the absence of a better plan, British Nuclear Fuels has been toying with the idea of postponing the decision: leaving the waste in domes just under the surface of the earth until someone in some future generation works out what to do with it.[85]

In the United States, workers at the agency responsible for testing the Yucca Mountain repository in Nevada, into which the federal government intends to dump all the nation's nuclear waste, falsified the rates at which water percolates through it. An employee of the US Geological Survey admitted that

I keep track of two sets of files, the ones that will keep QA [Quality Assurance] happy and the ones that were actually used.[86,87]

The purpose seems to have been to make the site seem safer than it is. It now looks as if Yucca Mountain is an unsuitable repository.[88]

The enormous cost of waste disposal and the decommissioning of power plants is one of the reasons why nuclear power keeps guzzling public money. In the United Kingdom, cleaning up our nuclear sites will cost £70 billion.[89] Because the anticipated price has risen steadily

---

* Much of the population around Sellafield depends on employment by the nuclear industry.

over the past ten years, it would be fairer to say 'at least £70 billion'. Even before this spending begins, the government has been quietly handing over our money. In 2002, it lent British Energy £650 million.[90] In 2005, a leaked document revealed that it had given the company a further £184 million.[91] The public was never officially informed.

Nowhere is there a nuclear power station which does not rely on subsidies of one kind or another. Even the famous Olkiluoto reactor in Finland, which is the only nuclear power station currently under construction in Europe, and the only one being built anywhere without government money, now seems to be a loss leader underwritten by the French company Areva, in order to create the impression that the technology is commercially viable.[92]

A further hidden subsidy, whose actual cost is impossible to account, is the insurance cover the state provides. The financial risk of a nuclear accident is so high that commercial insurers won't cover it. Three international treaties limit the nuclear operators' liability:* the state will pick up the bill instead. In the United Kingdom, the government will cover any accident costs greater than £140 million.[94] In the United States the figure is $200 million;[95] in Canada, a mere CAD$75 million.[96] But the European Parliament estimates that the cost of a large-scale nuclear accident ranges anywhere from €80 billion to €5.5 trillion.[97]

In 2005 the economic consultancy Oxera calculated that replacing the United Kingdom's nuclear power stations, most of which need to be retired by 2020, would cost around £8.6 billion, excluding insurance and other guarantees but including future decommissioning and waste disposal. About £1.6 billion of this would have to come from the government.[98]

Any electricity which is more expensive than the cheapest kind on the market is likely to need government support if its operator is not to go out of business. This applies to renewable power as much as to nuclear energy. But a study in the United States shows that during the first fifteen years of the development of the two industries, nuclear

---

* The Paris Convention (1960), the Vienna Convention (1963) and the Joint Protocol (1988).[93]

power received forty-four times as much government money as wind power.[99]

I think there are two reasons why governments have been so generous to nuclear power. The first is that it was used – especially, in 1953, by President Eisenhower – as a demonstration of the potential for disarmament. He believed that the nuclear sword could be beaten into the nuclear ploughshare.

It is not enough to take this weapon out of the hands of the soldiers. It must be put into the hands of those who will know how to strip its military casing and adapt it to the arts of peace.[100]

His programme ('Atoms for Peace'), which had the unintended consequence of equipping non-nuclear nations with the fissile materials they could use to make nuclear bombs, was enthusiastically adopted by the other members of the UN Security Council, perhaps because it afforded them a degree of political cover for their own expanding weapons programmes.

The second is a perverse effect I have noticed when investigating other development projects:[101] that big, expensive schemes often find more favour with governments than small, cheap ones. This is partly because a small number of large projects is easier to administer than a large number of small ones. But it is also because the bigger and more expensive a project is, the more powerful the lobby which demands that it be approved. Nuclear power stations can be built only by large construction companies, and large construction companies swing more weight with the government than the small operators hoping to install wind turbines.

So how much does it cost? The only honest reply is that I haven't the faintest idea. To explain why, the table below shows some estimates. They are all for the wholesale price of electricity from nuclear power. The average wholesale price of electricity at the end of 2005 was roughly 3.6 pence per kilowatt hour. This may be anomalous: at the end of 2004 it was 2.1 pence per kilowatt hour.[102]

source	price per kilowatt hour of electricity
Nuclear Energy Institute[103]	1.7 US cents (1.0 pence)
Royal Academy of Engineering[104]	2.3 pence
British Energy and British Nuclear Fuels[105]	2.5–3.0 pence
UK government (in 2020)[106]	3.0–4.0 pence
Massachusetts Institute of Technology[107]	7.0 US cents (4.0 pence)
New Economics Foundation[108]	3.4–8.3 pence

I conclude that the price of nuclear power is a function of your political position. If you don't like it, it is expensive. If you do like it, it is cheap. But perhaps there is an easier means of determining whether or not nuclear power is commercially viable. There is no law against building nuclear generators in the United Kingdom. We have a deregulated market in electricity, which encourages suppliers to find the cheapest means of producing it. So if – as the Nuclear Energy Institute suggests – nuclear power is cheaper than its competitors, you would expect companies to be replacing redundant plant of other kinds with atomic power stations. But the last one to be built in this country was Sizewell B, whose planning application was submitted in 1981, and whose construction commenced in 1988.[109]

Three more big questions about nuclear power remain. The first is whether there is sufficient uranium (the principal nuclear fuel) to keep the industry going. This is a difficult question to answer, because it depends on several unpredictable factors. One of them is the amount of money people are prepared to pay for it. This might sound odd, but it applies to every mineral: the more it is worth, the more there is. Seams which were previously too expensive to mine become exploitable as the price rises. Another is the level of geological knowledge: it is never easy to determine how much of the total global reserve has already been identified. A further factor, peculiar to the nuclear industry, is whether or not spent uranium fuel will be reprocessed and used in fast-breeder reactors. The answer to this question is that we should hope not, as these activities expose us to peculiar dangers. Fast-breeder reactors use more concentrated nuclear fuel, which means that accidents could be more dangerous than accidents in other kinds of fission reactors. Reprocessing, or so the perennial mishaps at Sellafield suggest, increases the spillage of radioactive materials into the environment. It also separates plutonium from the other wastes, providing greater opportunities for the proliferation of weapons or for seizure by terrorists.[110]

The World Nuclear Association claims that known reserves of uranium in the 'lower cost category' amount to around 3.1 million tonnes. At current rates of use, they will last for half a century. It points out that 'this represents a higher level of assured resources than is normal for most minerals.'[111] A widely circulated and detailed paper by the energy analysts Jan van Leeuwen and Philip Smith estimates that if all the world's electricity was produced by nuclear power plants, uranium supplies would last for 6.8 years.[112] But the British government's advisers, the Sustainable Development Commission, after examining the issue in great depth,[113] concluded that

On current predictions, there are no major concerns over the long-term availability of uranium . . .. in the past uranium reserves have been consistently underestimated . . . there is probably enough uranium at a reasonable price to match future demand.[114]

As this is an inconvenient finding for the Commission, which fiercely opposes nuclear power, I am inclined to trust it.

Uranium, like coal, is extracted in open-cast mines. Because there is less of it, and less fuel is used per watt of power, the mines take up a much smaller proportion of the planet's surface than coal quarries. But the spoil and tailings – the rocks left behind when the uranium has been extracted – are more toxic.[115]

Closely associated with this issue is the second big question: how much carbon dioxide is saved by using nuclear power? The reason these two issues are linked is that the poorer the quality of the uranium ore, the more energy is required to mine and process it.

The Sustainable Development Commission does not address the question about the impact of low-grade uranium on carbon emissions, perhaps because it believes it will not arise.[116] The World Nuclear Association claims that with uranium ore of the grades used today, nuclear power consumes about 1.7 per cent of the energy it produces.[117] This includes the energy costs of building and decommissioning the plant and disposing of the waste. If a 'very low grade ore' – containing just 0.01 per cent uranium – was used, the energy cost would rise to 2.9 per cent of the total output, because more energy would be required to separate the uranium from the ore.[118] If this is true, then all our electricity problems are solved: even with low-grade ores, we could cut our carbon emissions by 97 per cent. But the figures produced by the industry's critics are wildly different. Van Leeuwen and Smith maintain that using ores which contain 0.02 per cent of uranium or less consumes more energy than it produces. Their charts show a net energy production from ores containing 0.01 per cent uranium of between *minus* 200 per cent and *minus* 500 per cent.[119]

In both cases the figures look – to my layman's eyes – well-sourced. By this I mean that references are given, and those references lead back to real papers.[120,121] To claim to know which – if any – account to trust would be to feign an olympian knowledge I do not possess.

The third big question is this: assuming, for the sake of argument, that nuclear power really does provide electricity that is largely carbon-free, can it be delivered quickly enough to meet our target?

At first the answer appears to be 'no'. In his submission to the House of Commons Environmental Audit Committee, the environmental analyst Tom Burke argued that if the British government had decided to build nuclear power plants in 2005, the earliest date by which the first plant could be operating would be 2021.[122] The sixteen years would be needed to obtain finance and planning permission and to design and build the plant. Similarly, the British government found that if the Advanced Passive 1000 reactor (the most likely model) were to be used, a new nuclear programme would take at least twenty years to come to fruition.[123]

But if this demolishes the case for nuclear power, it also demolishes the case for offshore wind farms, new railway lines, better car engines and almost everything else we might seek to develop. If we are to install a large number of new wind farms, for example, we'll need some major new connections to the national grid. The Scottish and Southern Energy Group is currently trying to build a new power line across Scotland for this purpose. I asked it how long it thought this would take.

We began the process for this project in 2002. We applied for the consent in 2005. We anticipate that the work will take at least four summers. What we don't know is how long the consent-granting process will take; a public inquiry, should there be one, would obviously add a considerable length of time to this. In short, if we started next summer and all went to plan, the project would have taken seven years; but with uncertainty over the length of the consents process, it could be longer.[124]

According to Professor Nick Jenkins of Manchester University, a line of this kind can be expected to take ten years.[125] Alongside this process, of course, the company needs to obtain permission to build the wind farms, and this, being controversial, can also take years.

But when a country really wants something to happen, the usual constraints can be swept aside. In his book about the US automotive industry – *Taken for a Ride* – Jack Doyle shows how the car manufacturers responded to the bombing of Pearl Harbor.

From a standing start in late 1941, the automakers converted – in a matter of months, not years – more than 1,000 automobile plants across thirty-one states . . . In one year, General Motors developed, tooled, and completely built from scratch 1,000 Avenger and 1,000 Wildcat aircraft . . . GM also produced the amphibious 'duck' – a watertight steel hull enclosing a GM six-wheel, 2.5 ton truck that was adaptable to land or water. GM's duck 'was designed, tested, built, and off the line in ninety days' . . . Ford turned out one B-24 [a bomber] every 63 minutes . . . Barely a year after Pontiac received a Navy contract to build anti-shipping missiles, the company began delivering the completed product to carrier squadrons around the world.'[126]

If our governments decide that climate change is an issue as urgent as the international crisis in 1941 – in my view a reasonable comparison – they could turn the economy around on a sixpence. Planning objections would be ignored, incentives and regulations would be used to make companies move as swiftly as General Motors and Ford responded to the war. Wind farms, powerlines and nuclear power stations – if this is what we want – could all be built in much less than a decade.

With this in mind, I think we can reassess the International Energy Agency's pessimistic assumptions about carbon capture and storage.

While many of our existing power stations cannot be retro-fitted with the necessary technology, most of them are already on their way to the knacker's yard. In the United Kingdom, which seems to be fairly typical, we need to replace nearly 50 per cent of the electricity generating capacity we possessed in 2000 by 2018, and 90 per cent by 2030.[127] This is an important fact to remember throughout this discussion: we are not talking about demolishing useful plant, but about replacing power stations which are already becoming redundant. On the other hand, carbon-capture and storage technology – while demonstrated – is still immature. It has not yet been tested on the scale I am talking about. But if it can be shown to work in all cases, there is no good reason why every new power station burning fossil fuel that we built between now and 2030 could not be designed to remove and bury its carbon.

Despite all the uncertainties I have encountered, I think I have grounds for making my decision. Because of the industry's record of corner-cutting, because of its association with the proliferation of weapons of mass destruction and because of the unresolved questions about waste disposal and the energy balance, I will provisionally place nuclear power second from last in my list of preferences, just above generation using coal from open-cast mines. And I will propose carbon capture and storage as a partial solution to the problem. The current state of the technology and the replacement rate of power stations suggest that, with sufficient political will, gas-fired power stations fitted with carbon capture equipment could provide roughly 50 per cent of our grid-based electricity by 2030.

A greater contribution than this, however, is unlikely, so to reach my target I will have to look elsewhere. The obvious alternative is renewable energy. But how much of our electricity can it supply, and at what cost?

6

How Much Energy Can
Renewables Supply?

And tempest roars, with tempest vying,
From sea to land, from land to sea,
In their alternate furies tying
A chain of deepest potency

Faust, Prologue in Heaven[1]

There are many things I dislike about renewable energy, or – to be
more precise – about the industry that promotes it. I dislike the
misleading claims its advocates make. I dislike the tokenism that
attends it: a petrol company might put a wind turbine beside a filling
station (because it is too far from the grid to make a connection
worthwhile) and its customers think it has gone green. I dislike the
way in which covering the countryside with wind turbines is often
seen as a solution to our excessive consumption of fossil fuel, as if
the new technology, in the absence of policy, replaced rather than
simply augmented the old one. I dislike the way in which environmentalists sometimes choose to overlook the destructive effects of
renewable technologies, such as the tidal barrages which drown the
ecosystems of estuaries.

But even in my grumpiest moods, I can also see its virtues. The
wind, waves and sun are not going to run out – or not while we still
occupy the planet. Neither Mr Putin nor any other energy monopolist
can switch them off. No wind farm can ever melt down, or present a
useful target for terrorists (modern Don Quixotes notwithstanding).
Decommissioning is cheap and safe. The energy required to build the
machines on the market today is a small fraction of the energy they

will produce,[2,3] and as soon as that has been accounted for, they emit no carbon. While renewable technologies can dominate a landscape, this impact is surely less significant than the destruction of the biosphere. But even if we were to overlook its damaging or counter-productive effects, there remain good reasons for questioning the claim that our electricity could be supplied wholly or even largely by renewable power. As I hope you are discovering, nothing in this field is simple.

The first question that people who advocate the replacement of our power stations with renewable energy must answer is this: is there enough of it? The United Kingdom – islands surrounded by high winds and rough seas – has the best resources in Europe. But even here there is room for doubt about whether ambient power would be sufficient to meet our demands.

We use around 400 terawatt hours of electricity a year.[4] A terawatt hour (TWh) is energy produced at the rate of 1 trillion joules per second for one hour. That may not make it much more comprehensible, but you need bear only the comparative figures in mind. In 1999, a consultancy called the Energy Technology Support Unit was commissioned by the British government to work out how much renewable energy the country has. The table overleaf gives the figures it produced for what it calls the 'practicable resource'. Practicable resource means the power that could be produced for a reasonable cost after various constraints (such as not building in national parks or where the seabed is too soft, or where turbines would interfere with radar or migrating birds) have been taken into account. A 'reasonable cost' was 7 pence per kilowatt hour (kWh) or less. Keeping costs down to this level means that the machines can be built only where the resource is concentrated – in other words where the wind blows hard or the tide flows strongly – and only where they can be clustered. Scattered wind turbines, for example, would cost a fortune to connect to the grid.

You might notice that some technologies, such as tidal barrages (a dam across an estuary), are not listed in the table. This is because the consultancy concluded that the practicable resource is zero – something I am rather glad about. So, astonishingly, it looks as if even in the windiest, most battered nation in the European Union,

energy source	practicable resource in 2025 (terawatt hours)
onshore wind	8
offshore wind	100
waves	53
tidal stream[i]	2
solar photovoltaic	0.5[ii]
hydroelectric	7[iii]
Total	170.5

i. 'Tidal stream' means rotors making use of free-flowing current, as opposed to 'tidal barrage', which means a dam across an estuary.

ii. ETSU's report is extremely confusing on this point. It starts off by claiming that 'the maximum practicable resource in 2025 is 266TWh/year [TWh means terawatt hours]. The maximum practicable resource includes PV applied to all orientations.' This is – to say the least – an eccentric proposition: that solar panels can be fitted to building surfaces at all points of the compass. It then goes on to claim a 'technical potential' in 2025 of 37 TWh, and a 'market potential' of 0.17. Then it produces a 'resource-cost curve' showing how much electricity can be produced at 7p per kWh or less, which gives 0.5 TWh at an 8 per cent discount rate. This appears to be its final figure. The Royal Commission on Environmental Pollution seems to interpret its report as expressing pessimism of this order, when it says that 'ETSU estimates that there will be no photovoltaic resource available at a cost of less than 7 p/kWh by 2010, and only a very limited amount (average output of 0.2 GW) by 2025.' But I cannot find the 0.2GW figure (which is actually a measure of capacity, not output) in ETSU's report.

iii. 4TWh is already in operation. ETSU suggests that the only significant remaining resource whose exploitation would not cause serious problems is 3TWh in Scotland.

Source: Energy Technology Support Unit.[5]

we could harness less than half the ambient energy we'd need to produce our electricity.

It is now clear, however, that these estimates are far too pessimistic. The Energy Technology Support Unit (ETSU) assumed that the national grid remains unmodified – in truth it could be extended into new areas. The costs of some renewable technologies have already been falling much faster than it proposed. It investigated the potential for wave power in just five places: in reality we could use it over a much wider area.

In 2005 a government paper called *Offshore Renewables – The Potential Resource* produced a number for what it called the 'Potential offshore wind generation resource in proposed strategic regions'. This it estimated at 3,213 terawatt hours: over eight times our total electricity demand.[6] What makes it even more astonishing is that the 'proposed strategic regions' are all off the coast of England: the figure does not take account of the seas off Scotland, Northern Ireland and Wales, where the wind is generally stronger (but, because the water is mostly deeper and rougher around these countries than around England, it is harder to plant turbines there).

The paper does not explain how much of this 'potential resource' could become a 'practicable resource'. But a likely estimate is between 5 and 10 per cent. So if we could cut our demand by 25 per cent – reducing the 400 terawatt hours we use each year to 300 – offshore wind power alone could, in theory, supply somewhere between one half and all of our electricity.

One of the reasons why the government's new estimates are so much greater is that offshore wind turbines are already bigger than ETSU envisaged. ETSU suggested that each turbine would have a capacity of 1.5 megawatts, while the government's paper proposes 3 megawatts. In fact turbines of this size are already being installed. Some people are now suggesting that by 2030 10-megawatt machines could be built.[7] For the same reason, we can regard ETSU's figure for the onshore wind resource as too low, as it assumed that the machines erected on land would have a capacity of only 600 kilowatts.[8] Already 2-megawatt windmills are being built on some of our hills. This seems to be the maximum size for onshore turbines however, as there are no lorries big enough to carry longer blades.

ETSU also assumed that no wind turbine would be built in water deeper than 40 metres. The government believes that they could be planted in the bed of seas as deep as 50 metres. It also envisages that they could be built much further from the land. If you built a very large wind farm 100 kilometres, rather than 20 kilometres, from the shore, its paper shows, the extra expense of attaching it to the national grid would add only about 5 per cent to the total cost of the project:

'It is therefore likely that the industry could consider potential sites well outside territorial waters, and perhaps as far as 100 kilometres offshore.'[9]

Though it does not mention them by name, the government seems to have discovered something that has also captured my attention: High Voltage Direct Current cables. I am coming to believe that they could change the world.

The first electricity networks, built by Thomas Edison, carried direct current (DC) electrical power, but only at low voltages. This meant that they were restricted to short distances: a maximum of about two kilometres. Alternating current (AC) could be generated and distributed at high voltages, and soon became the means by which electricity was transmitted almost everywhere. The pylon lines upon which we all depend and which we all detest are – almost without exception – high voltage AC. But the development of the high-voltage valve and much lighter wires, which permit the long-distance transmission of direct current, changes the formula.

Setting up a high voltage DC line is, in the first instance, more expensive than setting up a high voltage AC line. The initial loss of electricity when you hoist it onto the transmission system is greater.[10] But the longer the transmission distance, the better DC begins to look. DC pylons are smaller, so the transmission network requires less land (and is not nearly as ugly). A graph published by the World Bank shows AC costs exceeding DC costs beyond about 650 kilometres.[11] This distance is falling rapidly, because the costs of DC systems are plunging, while the costs of AC systems (due to their higher environmental impact and requirement for land) are rising. The World Bank claims that new DC cables made of extruded polyethylene could now make economic sense over as little as 60 kilometres.[12] But most importantly, though the initial electricity loss on a DC line is higher, it does not increase with distance. On AC systems, by contrast, the longer the line, the more you lose. There is no inherent limit on the length of a DC cable. Already there is a line in the Democratic Republic of Congo that is 1,700 kilometres long.[13]

What this means is that you can draw your electricity from a far greater area than before. High voltage DC, which can be run along the sea bed, opens up any patch of sea shallower than 50 metres to wind turbines, and pretty well all the continental shelf to wave power devices, which (because they float) can be anchored at greater depths. Since wind speeds rise by around one metre per second with every 100 kilometres from the shore,[14] this means that the cost of renewable power could actually fall with distance from the coast.

While offshore wind power is currently more expensive than on-shore wind, the economies of scale permitted by massive develop-ments attached to long-distance cables means its price could fall very rapidly: according to one estimate, by 40 per cent in ten years.[15] And, beyond a certain distance, the only people whose aesthetic sensibilities are likely to be offended are trawlermen. You can install wind turbines which rotate faster (and are therefore both noisier and more efficient) without upsetting anyone.[16]

But it's not just new wind and wave power that the long lines could exploit. At the moment there is an inverse relationship between the availability of solar power and human habitation. It is most concen-trated and most reliable in deserts. For years, rogue environmentalists have been pointing out that solar electricity generated in the Sahara could supply all of Europe, the Gobi could power China, and the Chihuahuan, Sonoran, Atacama and Great Victoria deserts could electrify their entire continents. These people have been dismissed as nutters. The development of cheap DC cables suggests that they might one day be proved right.

There are several technologies with the potential to produce cheap electricity in very sunny places. One involves the mass production of solar photovoltaic cells of the kind now being installed, in small panels, on people's roofs. On a small scale, at high latitudes, solar electricity is – and will probably remain – more expensive than almost any other kind of electricity. But huge photovoltaic farms in the desert have the potential to realize economies of scale which have hitherto been unavailable. Perhaps even more promising is a technology called 'solar thermal electricity'.

You can use a reflective dish or trough to focus heat from the sun onto a tube containing water or another liquid. This reaches a temperature of about 400°, and the steam drives an engine which generates electricity.[17] Power farms using this technology have been operating in southern California since the early 1980s. Their output costs between 12 and 15 US cents (about 7–9 pence) per kilowatt hour.[18] At times of maximum demand in southern California, electricity now costs between 10 and 18 cents.[19] (Because of air conditioning, demand is greatest there when the sun is hottest, so the maximum output of solar electricity happens to match the maximum demand.) There are now similar schemes in Spain, Italy, Morocco, India, Mexico and several other countries.

Or you can use mirrors to bounce the sunlight onto a central receiver. The mirrors track the sun and focus its heat onto a small ceramic plate on a tower about 100 metres high. The plate reaches a temperature of 1000°, again generating steam which drives a turbine.[20] A demonstration plant is already working in Spain. With temperatures like this, it should be possible to store some of the energy the plant produces in the form of molten salt, which would allow it to generate electricity even when the sun is not shining.[21] Another scheme, which has also been tried at an experimental level in Spain, uses heat from the sun to drive air through turbines in a hollow tower. The Australian government has shown great enthusiasm for a solar 'power tower' 1 kilometre high in the outback north of Melbourne,[22] but I suspect its interest is prompted more by the prospect of building the world's tallest structure than by the project's commercial viability, which is currently not demonstrable.

The International Energy Agency calculates that if solar photovoltaic panels were used to cover 50 per cent of the land surface of the world's major deserts, they would produce eighteen times as much energy – which means 216 times as much electricity* – as the world now uses.[25] In other words, you would need to cover only 0.23 per cent of the land to meet demand, assuming that people used electricity only when the sun was shining. In 2003 the agency suggested that this electricity would cost between 9 and 11 US cents per kilowatt hour, and that this would fall by between a quarter and a half by 2010.[26]

What makes the use of long-distance cables particularly exciting is that they allow you to obtain electricity from several different regions

---

* IEA figures suggest that the world uses 200,000TWh of energy,[23] of which 16,742TWh is electricity.[24]

at the same time. This makes renewable energy much more reliable than it would otherwise have been. There might be a flat calm in the North Sea, but a steady breeze 100 miles off the coast of Ireland. When the sun in the eastern Sahara is going down, the sun in the western Sahara is at full strength. Long-distance cables, as well as opening up new sources of electricity, might also help to over-come the greatest constraint limiting the use of ambient power – its intermittency.

I have mentioned the reductions in cost which could be achieved as a result of economies of scale. But in another respect renewable power behaves very strangely: the more there is, the more expensive one of its components becomes.

The problem comes down to this: that the wind does not blow and the waves do not rise all the time. Unlike power stations which burn fossil fuel, they cannot be turned on when we want them. If we switched our entire electricity generating network over to variable sources of renewable power, without building a massive energy stor-age system, then whenever the wind or waves dropped, the grid would collapse. What this means is that our renewable energy needs to be supported by other forms of power.

All electricity generating systems have spare capacity. When a steam turbine trips out or a transformer fails, or there's a sudden and unexpected surge in demand, the extra plant, which was either lying idle or operating at half-throttle, can be fired up to fill the gap. In the United Kingdom, for example, where the maximum demand for electricity requires 62.7GW of generating capacity, we have 75.5GW of plant.[27] (Only some of this spare capacity can be used, however, as it includes generators that are out of service.)

The variable nature of wind and waves and other renewables means that though we might build many gigawatts of renewable generating capacity, we cannot retire a corresponding amount of power stations that burn fossil fuel. We will have to carry the cost of maintaining them and even, in the future – when the old plants die – building new ones, whose main purpose would be to sit and wait for the wind to fail. The more renewable installations we build, the greater is the impact on the grid of the wind ceasing to blow or the waves ceasing to rise. So each additional gigawatt of renewable power we install displaces a smaller proportion of conventional plant.

If we built 8 GW of wind farms, for example, they would allow us to shut down around 3 GW of power stations which burn coal or gas.[28] But if we built 25 GW of wind (which would produce about 20 per cent of our electricity), we could shut down only 5 GW of old plant.[29] Beyond that point, very few old power stations could be closed, however many windmills we built.[30] Graham Sinden of Oxford's Environmental Change Institute has shown that a more reliable mixture – of wind, waves and tidal power, rather than just wind – would allow us to raise the retirement rate a little: 26 GW of renewables could shut down about 6 GW of old plant.[31] Even so, this means we would have to sustain 96 GW of power generation in order to provide the electricity we now obtain from 76.* It's important to note, though, that while the capital costs of the electricity generating network would rise, the cost of buying fuel would fall, as the gas or coal burning plants providing back-up power for the new wind farms would be used much less often that they are today. As long as wind power pays its way, the fact that our total generating capacity is bigger doesn't matter.

A report commissioned by the British government shows that if we obtained 20 per cent of our electricity from renewables in 2020, the extra 'system costs' would add between £140 million and £400 million to a total generating cost of £9 billion. If renewables supplied 30 per cent of our electricity, the system costs would amount to an extra £330–920 million.[32] The great majority of these extra costs would be for 'balancing and capacity', which means keeping the old plants alive and firing them up when they're needed. Even so, as a report by the UK Energy Research Centre suggests, if 20% of our power came from renewables, the extra system costs of a massive increase in the number of wind turbines are likely to be far lower than the possible economies of scale.[33]

And beyond 20 or 30 per cent? Well, hardly anyone even seems to be asking the question. I still haven't got to the bottom of why this is. The Royal Commission on Environmental Pollution complains

---

* 26 – 6 = 20; 20 + 76 = 96.

that 'there appears to have been no research as yet' into the question of how much renewable electricity the grid can take, 'either by the National Grid or by any other body'.[34]

I suspect that researchers won't go beyond 20 per cent or 30 per cent because no one is asking them to do so. The academic study of carbon reduction in the rich nations is blighted by government policy. Governments (which provide most of the money for research) set targets (in the UK's case, 20 per cent renewables by 2020) and then commission people to find out how they could best be met. They do not commission them to find out whether they are the right targets, or whether they bear any relation to the technical and economic limits. So beyond 20 or 30 per cent, we are groping about in the dark.

Building more wind turbines and wave machines and other renewable generators will not allow us to shut down most of our existing power stations. But this does not mean that our remaining thermal plants will be burning as much fossil fuel as they do today. Those which are retained for the purposes of insurance will, for the most part, be fired up only when demand is high and the wind starts to drop.

The power stations which respond fastest or cost least to start either burn coal or burn gas in an 'open cycle' turbine.[35] These are both inefficient means of generating electricity, so the carbon cost of wind is a little higher than it might otherwise have been. Moreover, more plant than before will have to be kept 'part-loaded'. This means it is turning over at reduced capacity but ready to be taken up to full power almost instantly. Power stations running at part-load are 10–20 per cent less efficient than power stations running at full load.[36]

According to the Royal Commission on Environmental Pollution,

Although the total capacity of those [backup] plants would be substantial, it would be used only infrequently, and the resulting addition to the annual total of UK carbon dioxide emissions would not therefore be large.[37]

But because it does not consider any contribution from renewables beyond 20 per cent, we don't know whether this would still be the case if we started producing, say, 50 per cent of our electricity by these means. A paper published in the journal *Energy Policy* goes into a little more detail:

Taking a conservative estimate of 10 per cent for the reduced efficiency [of part-loaded power stations] . . . the emission savings from the wind will be reduced by a little over 1 per cent. This can be compared with the 20 per cent of fossil fuel avoided by using wind generation.[38]

In other words, almost 99 per cent of the electricity produced by wind power would be carbon-free. But it also leaves us in the dark if we want to know what happens beyond 20 per cent.

Is this assessment reliable? Writing in the magazine *Civil Engineering*, the energy consultant Hugh Sharman points out that forecasters are currently unable to predict wind speeds on the following day to a higher degree of accuracy than 1.5 metres per second.[39] Whether the wind is blowing at 7.5 metres per second or 9 makes a big difference to the output of a turbine: Sharman suggests that the difference corresponds to 21 per cent of its total power-generating capacity. The wind's unpredictability he says, means that a large number of conventional power stations need to be kept part-loaded, or the national grid would become unmanageable. As a result, it would not be sensible for the UK to build more than 10 gigawatts of wind turbines.[40]

I have looked into his claims in some detail, and found that they cannot possibly be correct. The output of a single turbine – which might indeed be quite erratic – bears little relation to the output of all the turbines on an electricity grid, which fluctuates less rapidly. He has chosen the most sensitive portion of a wind turbine's output: at greater or lesser windspeeds, a change of 1.5 metres per second makes much less difference to the power it produces. More importantly, wind forecasts made one day ahead have little bearing on balancing an electricity grid, which takes place over a much shorter timescale. In the UK, forecasts are made for the following hour, and the errors are generally very small.

The cost of electricity from wind farms, like the cost of all other forms of power, depends on who has performed the calculation: the question has become as much political as economic. The table below gives some estimates. They take into account all the costs of wind generation: buying the turbines, erecting them, connecting them to the grid and providing backup power. The average wholesale price of electricity from conventional power stations was 2.1 pence at the end of 2004 and 3.6 pence at the end of 2005.[41]

source	onshore wind (per kilowatt hour)	offshore wind (per kilowatt hour)
Performance and Innovation Unit, 10 Downing Street – in 2020[42]	1.5–2.5 pence	2.0–3.0 pence
International Energy Agency – present day[43]	3.0–7.0 US cents (1.7–4.0 pence)	7.0–12.0 US cents (4.0–7.0 pence)
Performance and Innovation Unit, 10 Downing Street – present day[44]	2.5–3.0 pence	5.0–6.0 pence
Sustainable Development Commission – present day[45]	3.2 pence	5.5 pence
Royal Academy of Engineering – 'the future'[i][46]	4.8 pence	6.3 pence
Royal Academy of Engineering – present day[47]	5.4 pence	7.2 pence

i. It doesn't say when.

The British prime minister's office predicts that in 2020, electricity from gas will cost between 2.0 and 2.3 pence, while electricity from coal will cost 3.0–3.5 pence.[48] These might be low estimates, as the wholesale prices have greatly increased since it published its report in 2002. Either way, if its figures are to be believed, onshore wind will provide the cheapest form of electricity in 2020, while offshore wind will be broadly competitive with conventional generation. The academics I have spoken to maintain that the Royal Academy's figures are not supported by other studies and are considered by most analysts to be greatly inflated.

Though offshore turbines produce electricity more consistently than onshore machines, they are more expensive to build. Planting one in the seabed costs more than erecting it on land. The electronic equipment needs to be protected from the salt, and submarine cables are more expensive to lay than overhead power lines on land. But their costs will fall more rapidly than those of onshore windmills.[49]

Wave and tidal power are likely to remain more expensive than wind, as the table overleaf shows.

source	wave (per kilowatt hour)	tidal (per kilowatt hour)
Performance and Innovation Unit, 10 Downing Street – in 2020[50]	3.0–6.0 pence	
Jake Chapman and Robert Gross – by 2012[51]	4.0–5.0 pence	4.0–5.0 pence
Jake Chapman and Robert Gross – 'short term'[52]	4.5–6.0 pence	4.5–6.0 pence
Performance and Innovation Unit, 10 Downing Street – 'the first commercial-scale device'[53]	4.0–8.0 pence	4.0–8.0 pence
Royal Academy of Engineering – 'the future'[i54]	5.7 pence	5.7 pence
Royal Academy of Engineering – present day[55]	6.6 pence	6.6 pence

i. The Academy has not included the costs of standby generation for these technologies, because it does not expect much wave or tidal power to be used.

In one respect, the extra costs that renewable electricity might invoke have been exaggerated. The United Kingdom, like every developed nation, has a vast reserve of standby power that is, and will always be, maintained, and which is seldom included in the official figures: the emergency diesel generators owned by hospitals, army barracks, police stations, airports, offices and factories.[56] They retain them in case the grid breaks down. Altogether there is somewhere between 12GW[57] and 20GW[58] of this invisible reserve: in other words, we have 16–26 per cent more generating capacity than the government's numbers suggest.* If the higher figure is correct, and this reserve could all be called upon when renewable power failed, we would already have enough to support a 20 per cent contribution from wind, waves and tidal power of the kind described by Graham Sinden.[59]

National Grid Transco, the company which keeps the electricity on the wires, already uses a small amount of this hidden capacity. It has a 'standing reserve' agreement with a few of the people who own these generators, under which they must start contributing to the grid within 20 minutes of receiving its signal.[60] This is quite easily achieved, as a diesel generator can reach full power from a cold start within 20 seconds. Wessex Water reports that it takes part in the scheme 'primarily because experience has shown that standby generators

---

* 87.5 or 95.5GW, rather than 75.5.

won't work reliably in an emergency unless they are full-load tested at least once a month'.[61] There is, as far as I can see, no theoretical reason why the rest of the country's emergency generators cannot be recruited for the same purpose. Because they already exist, and must continue to exist for other reasons, their use could greatly reduce the capital cost of standby power. The disadvantage of using these generators is that they have a lower fuel efficiency than large power stations, though this is offset somewhat by the fact that they respond faster. As they are likely to be used only for short periods, this use of fuel is not significant.

Another means of bringing down the cost is to make ambient energy more reliable. This sounds ridiculous: either the wind is blowing or it isn't. But that is generally the case only in one place. The greater the number of regions in which windmills are built, the higher the chances that some of them will be turning. A study by the consultancy Oxera of fewer than half the possible wind-generating regions in the UK* discovered that in any one year there are, on average, only 23 hours in which electricity demand is high and wind turbines would be producing less than 10 per cent of their maximum output.[62] Graham Sinden has studied the weather of the entire United Kingdom: 'Between 1970 and 2003, there was not an hour, let alone a day or a week, with no wind across the UK.'[63]

When wind farms are 1,000 kilometres apart, their output is correlated at just 10 per cent: in other words, there is a 90 per cent chance that wind speeds will not be the same in both places.[64]

One of the advantages of high voltage DC cables is that they could reduce the amount of generating plant required to insure against renewable electricity failures, as they greatly expand the number of places in which wind and wave energy could be captured. Sinden's work, as I've mentioned, also shows that renewable energy is more reliable if it does not come from just one source. The tides run whether or not the wind is blowing, and waves keep rolling across the sea long after it has dropped.†

---

* It considered offshore wind in the Wash, the Thames Estuary and the North West, and onshore wind in Scotland.

† Sinden shows there is a correlation of just 42 per cent between the power output of wind and the power output of waves.[65]

But as the amount of renewable power increases, and its reliability improves, we encounter the opposite danger: that of an embarrassment of riches. If the wind is blowing strongly while demand is low, then turbines will have to be shut down if the frequency of the alternating current on the grid (assuming we are still using an AC grid) is not to rise beyond its limits. This seems like a painful waste of energy. But there may be a way of solving two problems at once.

At the moment there is only one means of reasonably efficient long-term electricity storage, and that is the process I mentioned in Chapter 5: pumping water from a low reservoir up to a high one and leaving it there until there's a surge in demand. The energy losses involved in 'pumped storage' are about 20–25 per cent,[66] which compare to losses of 60 per cent or more from the possible alternatives.*

The reservoir at Dinorwig in north Wales cost, at today's prices, £1.6 billion to build.[67] But it has a very long life, so the annual cost is actually quite low. If it were possible to build new pumped storage systems to soak up surplus power from renewable generators, their running costs would be small, as the electricity – which would otherwise have been wasted – is effectively free. They could also solve many of the problems associated with the intermittency of wind and waves and tides: when the wind drops, the gates can be opened. And in this case there would be no carbon costs.

So can more pumped storage plants be built? The House of Lords says that 'the scope for increasing the volume of pumped hydro in the United Kingdom is limited by the same factors that limit conventional hydro.'[68] But I cannot see how this could possibly be true. Conventional hydroelectricity is limited by the availability of rivers flowing rapidly downhill. But you don't need a river to build a pumped storage plant: just a dip on the side of a mountain and another one at the bottom. By building small dams across both of them, you create a hydro system from scratch. It is easier still if one of the dips is

---

* Such as compressing air in underground caverns and the electrolysis of water to produce hydrogen. The other kinds of possible storage, such as flow batteries, flywheels or superconducting magnetic energy, tend to be useful only for much shorter periods.

already filled with water: a natural phenomenon commonly described as a lake.

The Royal Commission on Environmental Pollution raises a more plausible objection: 'because of the effects . . . on landscape and wildlife, it is unlikely acceptable sites could be found.'[69] This problem needs to be taken seriously. The dams and cables would certainly spoil the view. But given that many of our mountainsides are severely degraded by overgrazing, I am not convinced that the impact on wildlife will always be very great.

The attraction is obvious to anyone who studies a wind map of these islands. The greatest potential for generating power is off the coast of north-west Scotland. North-west Scotland happens to contain a large number of mountains, some of them close to the places in which the cables would come ashore. This means that scarcely any extra transmission networks would be required. But because there has so far been no public discussion of this idea, and no consideration of which mountains might be suitable and what the impacts would be, I am not putting it forward as a firm proposal. We should, however, start discussing it.

There is a further means of increasing the reliability of the system, and this is to alter the way in which people use electricity. If electricity use could somehow decrease when the wind drops, this would reduce the need for backup power, and the risk of the grid failing. But how on earth could that be done?

Several authors have suggested that appliances whose power does not need to be on all the time – such as fridges – should be designed to disconnect themselves when total demand rises.[70,71,72] When the lights come on – on a winter evening for example – fridges and washing machines should turn themselves off. They could respond automatically to changes in the frequency of electricity. If the frequency dips (because there isn't enough power on the grid), they switch off. If it rises, they turn themselves back on. This would reduce the great peaks in demand which force the electricity companies to keep so much plant on standby. It could also be used to help respond to the fluctuating levels of renewable energy. When demand is high and the wind is low, some of our appliances could switch themselves off. When demand is low and the wind is high, storage heaters, battery chargers and electrolysers could switch themselves on. This could answer one of the charges the critics of wind power make: that when the wind is blowing strongly, electricity will be wasted.

But there is a problem which most of the people who have written about 'demand management' have not addressed. The national grid company, sensibly enough, increases the frequency when it anticipates peaks in demand.[73] It will do the same if it sees that supplies of renewable power are about to drop. So our future smart fridges or smart washing machines would turn themselves on just when they should be turning themselves off. But it's not insuperable: the independent thinker Oliver Tickell seems to have solved it. Under his proposal – called the 'Real-Time Pricing Initiative' – the grid would double up as a communications network.

An open standard protocol [should] be developed and published to allow information about the instantaneous price of electricity to be broadcast through the electricity supply system. Much of the electromagnetic frequency spectrum would be available for this purpose without interfering with the principal power supply function, much as a telephone line can simultaneously provide voice and broadband internet service.[74]

'Smart plugs' attached to our fridges or washing machines would receive a signal that the marginal price of electricity is rising and switch their machines off, restoring power as the price falls back (or, in the case of fridges, before the temperature rises too much). He argues that

All the technologies involved are mature and available. All that is needed is good industrial design, and high production volumes.[75]

By altering the pattern of our demand, in other words, you can, in effect, improve the reliability of ambient power.

I am going to take four guesses, which appear to be supported by the evidence I have seen. The first is that the United Kingdom has sufficient renewable power comfortably to supply an average of 50 per cent of our electricity. The second is that the grid, and the reliability of the electricity it carries, could survive if 50 per cent of the supply came from renewables. The third is that the carbon costs of generating it would be considerably smaller than the carbon savings. The fourth is that the price per kilowatt hour would be no more than double the price the British government currently proposes for wind power supplying 20 per cent of our electricity.

In other words, my meta-guess is that 50 per cent is within the realm of feasibility. Depending on the cost of competing fuels, it might raise the price of electricity, but not grotesquely. I believe that the remaining 50 per cent, if we pursued a greatly accelerated development programme of the kind I discussed at the end of Chapter 5, could be supplied by thermal power plants whose carbon emissions are captured and stored. In other words, if my guesses are correct, all our electricity could be produced by two kinds of low-carbon generators: power stations burning gas whose exhausts are stripped of carbon dioxide, and renewable power plants, stationed either on our own soil or hundreds of kilometres away, and connected to the grid by means of long-distance cables. An electricity system running entirely on these two kinds of power (and conventional generators fired up to meet shortfalls in supply) would produce no more than 15 per cent of the carbon emissions currently released by our electricity suppliers. In combination with the efficiencies I discussed in Chapter 4, this would achieve an overall reduction of almost 90 per cent.

But I am sorry to say that, ambitious as this proposal is, it solves only part of the problem.

The reason is this: 82 per cent of the energy we use in our homes powers our *heating* systems (for both space and water heating), and only some of this heat is supplied by electricity: 17 million of our 24.5 million homes have gas-fired boilers.[76] Unless there is a means of solving the heating problem as well as the electricity problem, I will have to conclude that our homes – and our offices, factories, schools and hospitals – are unreformable. My targets, in that case, could not be met, and – given that we cannot do without heat – this project will have failed. I have somehow to find the means of heating our buildings without gas or coal. Because better insulation will give us a maximum likely carbon cut of around 40 per cent by 2030,[77] at least 50 per cent of the carbon reduction will have to be found by changing the way we generate heat. Is this possible?

Heating, at present, accounts for about 24 per cent of our

economy's entire energy consumption.[78] Altogether, we burn 2.4 exajoules of energy to produce the heat we need.[79] An exajoule is a marvellous figure. It is a million million million (otherwise known as a quintillion) joules. Around 70 per cent of these exajoules are used in our homes.[80]

At first sight, the solution seems obvious: we should do what humans and proto-humans have been doing for a million years or so, and burn wood. Trees absorb carbon dioxide as they grow, so as long as you harvest them at or below the growth rate you will produce no more carbon dioxide than they consume. In Scandinavia, which has, so to speak, more trees than you could shake a stick at, this is a viable solution. Wood already provides 17 per cent of all the energy used in Sweden and 20 per cent of the energy used in Finland.[81] But in the sparsely forested nations, it's a rather tougher target.

In its thorough report *Biomass as a Renewable Energy Source*, the Royal Commission on Environmental Pollution estimates that the calorific value of wood is around 10 gigajoules per tonne, and that the most efficient means of producing it – growing willow trees and harvesting their branches every three years or so – produces about 10 dry tonnes per hectare per year.*[82] Let us assume that growing, harvesting and transporting it uses 10 per cent of its energy content, and that the wood can be converted to useful heat at a rate of 75 per cent. Every hectare of land could then produce 67.5GJ of carbon-free heat. A gigajoule is a billionth of an exajoule. So to meet my target of 50 per cent of the heat we use – or 1.2 exajoules of zero-carbon energy – we would need 17.8 million hectares of land. The United Kingdom contains 17 million hectares of agricultural land, so we could just about accommodate it. You've probably spotted the flaw: we couldn't grow anything else.

This only begins to describe the problem. Already, during the summer, large parts of the United Kingdom, especially its agricultural land, are becoming subject to water stress, partly because of over-abstraction and partly because of climate change. In another report,

---

* There is a massive discrepancy between the figures it uses and those produced by the UN Food and Agriculture Organisation, which suggests a calorific value for willow branches of 18.4GJ/tonne.[83] I will stick for now with the Royal Commission's figures.

the Royal Commission notes that, 'Because they are fast-growing, energy crops need more water than arable crops.'[84] If we, and nations like ours, start devoting large areas of land to growing fuel rather than food, world food prices could start to rise, pushing malnourished people further towards starvation. In the longer term, the effect could be even more serious. In his book *When The Rivers Run Dry*, Fred Pearce has shown that falling water tables could threaten the world with famine.[85] Energy crops like willow trees accelerate the process by two means: partly by ensuring that the water tables fall faster than before, and partly by keeping land out of food production as food becomes scarce. The danger is that having built the infrastructure associated with a major investment in fuel crops, we will be reluctant to switch back to food production when it becomes a moral necessity.

Growing wood for heating isn't quite as pregnant with moral hazard as growing crops for road transport (which I will discuss in Chapter 8), for three reasons. The first is that woody plants can be raised on poorer land than the oil or sugar crops required by cars. The second is that growing wood for burning is a more efficient means of saving carbon dioxide. According to research conducted at Sheffield Hallam University, £1 spent producing bio-diesel saves between 3 and 6 kilograms of carbon, while one pound spent producing electricity from fast-growing trees saves 20 kilograms.[86] Growing it for heat, where conversion efficiencies are higher, is likely to save even more. The third is that the value per hectare of wood production is lower than the value per hectare of oil crop production, so farmers growing wood would have a greater incentive to return to growing food if the world is faced with a shortage.

Even so, it seems to me that their contribution to water stress and the upward pressure on the food price means we should restrict the cultivation of energy crops to a maximum of 20 per cent of our land area. This would allow us to produce 0.23 exajoules of carbon-free heating, or 19 per cent of the 1.2 exajoules I'm seeking.

We could, of course, increase this volume by importing wood from more forested countries. But biomass takes up a great deal of space (a cubic metre of dry woodchips weighs 150 kilograms, while a cubic metre of coal weighs about 800 kilograms[87]) so it costs a lot to transport. Because its energy density per cubic metre is low, the carbon costs of trucking it can swiftly start to counteract the carbon savings of using it.

If it were burnt on the farm, the total emissions from using wood to produce heat would be just 5 or 10 per cent of those caused by burning fossil fuel.[88] But for every 10 kilometres the fuel travels by road, the House of Lords calculates, 0.2 per cent of its energy value is consumed.[89] If it travels 1,000 kilometres, the net energy saving falls by 20 per cent. This is a worse deal than the raw figures suggest, for you are swapping transport fuels – which are likely soon to become scarce – for heating fuels, which will remain abundant. Unfortunately, in places like northern Russia, where there are few restrictions on the destruction of forests, wood is much cheaper than it is in western Europe, so it makes economic – if not environmental – sense to truck or ship it over here. And its importation cannot currently be prevented. As the British minister for science and technology pointed out, 'any restriction on fuel would not be permissible under international trade rules.'[90]

To the wood we grow, we could add the brashings, sneddings and sawdust produced by our foresters and the people who maintain our parks and gardens. This adds up, once other uses have been taken into account, to about 1.3 million dry tonnes a year,[91] or another 0.7 per cent of my requirement.*

Our farmers produce almost 4 million tonnes of surplus straw every year.[92] This will fall slightly if 20 per cent of our farmland is used for growing wood. The UN Food and Agriculture Organization, whose energy figures differ from the Royal Commission's, estimates that wheat straw contains about 93 per cent of the energy per tonne in willow branches.[93] For the sake of consistency, I'll apply that percentage to the Commission's figure, and give straw a value of 9.3 gigajoules per tonne. This provides 0.019 exajoules of carbon-free heat from 3 million tonnes of straw, or a further 1.6 per cent of the total.

The Energy Technology Support Unit also proposes that the United Kingdom could burn 1.3 million tonnes of chicken litter, which contain 13.5 gigajoules of energy per tonne, and about 1.8 million tonnes of 'animal slurries' (farmyard manure), whose energy content it does

---

* 6.75GJ (one tonne) × 1.3 million = 8.775 petajoules, or 0.0088 exajoules.

not mention.[94] But this suggestion disturbs me: it deprives the fields of manure, which has two major consequences. One is that more nitrogen fertilisers, which demand a good deal of fossil fuel to manufacture, will need to be produced; the other is that it will accelerate what could be an eventual global shortage of phosphate.[95]

So far, then, I have found 21.3 per cent of the 1.2 exajoules I am looking for. This assumes that the use of these biofuels is both straightforward and cost-effective. Neither assumption is entirely safe. If it turns out that the Food and Agriculture Organization's figures are more reliable than the Royal Commission's, we can raise this by a factor of 1.8, to 38 per cent. But you won't be surprised to hear me say that I don't know which numbers to trust.

The second most obvious source of carbon-free heat is the sun. The principle is simple: you place a panel of pipes resting on a black plate on your roof. The plate absorbs heat from the sun, warming the water in the pipes. The carbon costs of the system are approximately zero.

The consultancy company AEA Technology estimates that 50 per cent of the homes in the UK are 'physically capable of accepting a solar water-heating system'.[96] A report for the government by researchers at Imperial College suggests that if 50 per cent of our homes were fitted with solar heaters, they would produce 0.056 exajoules of heat, which is 4.7 per cent of our target figure. If the same ratio of heat production could be applied to our other buildings, the total contribution would be about 6.7 per cent. Unfortunately, solar hot water in the UK is, according to AEA Technology, 'much more expensive than heat from fossil sources', as the capital costs of installing the system are high, and the sun here is weak.

I am running out of options. Due to a regrettable absence of vulcanism, we have very few geothermal aquifers: bodies of hot water lying under the ground. The British Geological Survey says there are about 55 gigajoules of energy in aquifers with temperatures worth exploiting (40° or more). This means that, if it were not in the wrong place, you could mine it at the rate of 2.75 gigajoules per year for 20 years.[97] This is a long-winded way of writing zero.

'Ground-sourced heat pumps' are more promising. Below 1.5 metres, the earth has a constant temperature of 12°. If you either drill a borehole under your house, or run a zigzag of pipes under the soil in your garden, you can draw the heat from the earth and concentrate it to about 50°. This is just about right for underfloor heating, though a bit too cool for radiators. As the heat pump works by circulating water through the pipes, you need an electric motor to drive it. But the system generates between 2.5 and 4 times as much energy as it uses.[98] If the electricity comes from renewable sources, heat from the ground would be more or less carbon-free.

In the United Kingdom, we've been remarkably slow to use this technology. There are 230,000 ground-sourced heat pumps in Sweden and 600,000 in the United States.[99] Here there's a total of 300.[100] One problem is that our housing stock is replaced so slowly. Sinking a borehole or digging trenches then laying pipes under the floor is more easily done when a house is being built than when it's standing. Another is that we don't have much space. AEA Technology assumes that ground-sourced heat pumps can be installed either in new homes or homes in the countryside which are not attached to the gas grid (of which there are 4.4 million). If heat pumps were installed in *all* the homes that could take them, they could provide a total of 79.3 terawatt hours of heat per year, which equates to 0.022 exajoules, or another 1.8 per cent of the number I'm chasing.*

In commercial or industrial buildings there's the potential for a further 1.4 TWh per year, giving me another 0.033 per cent. I'm not doing very well.

All that remains is biogas: the methane produced by dumps, sewage farms and manure pits. Again, I am distressed to discover, the potential is small or less than small. AEA Technology says that constraints such as environmental hazard, public acceptance and the remote location of sewage farms (something for which in all other respects we are grateful) 'will make it almost impossible' for sewage gas to be used as a source of energy in buildings. 'The effective heat market is almost certainly zero.'[101] Gas from landfill sites (the term we now use to describe rubbish dumps) is subject to the same constraints: they are usually a long way from people's homes, so it is expensive to pipe the heat to where it's needed. On the whole, biogas is easier to use for generating electricity than generating heat.

---

* 1kWh = 3.6 megajoules; 79.3TWh = 22,028 terajoules, or 0.022 exajoules.

And that, as far as I can tell, is everything unless we are greatly to increase our electricity supplies and persuade people to return to electrical resistance heating, which is slow, inefficient and expensive. Even if I use the most generous figures and ignore some of the constraints imposed by cost, I have located only 46.5 per cent of the 1.2 exajoules of heat I was seeking.

But before I give up and become an aromatherapist – which appears to be the dreadful fate of all disillusioned activists – I have one last throw of the dice.

7

The Energy Internet

From your hearth the nimble flame . . .
Whence relief and comfort came.

Faust, Part II, Act V[1]

My final hope rests on this proposition: that I have been looking at the problem the wrong way round. I have been thinking about electricity and heating fuel as almost everyone has been thinking about them since construction of the national grid began in 1926: as commodities supplied over great distances from major sources. But there is an entirely different way of responding to the question of how our energy might be generated. It's generally described as 'micro generation' or 'the energy internet'.[2] In its pure form, it involves scrapping the national grid.

Instead of producing a large amount of power in a few places, the energy internet produces small amounts of power everywhere. Instead of using long-distance transmission cables to deliver it, it merely links up hundreds of micro generators in a local distribution web. With the help of a new kind of energy company, people buy electricity and heat either from tiny power stations built to serve their housing estates or, in effect, from each other. The local web should be more or less self-sufficient, but linked to other local webs to enhance its security.[3]

Greenpeace, which has become a powerful advocate of the energy internet, describes it thus:

Buildings, instead of being passive consumers of energy, would become power stations, constituent parts of local energy networks ... In

Greenpeace's vision ... coal-fired power stations have been closed and their surrounding webs of pylons dismantled, restoring swathes of countryside ... nearly all the input energy is put to use – not just a fraction as with traditional, centralized fossil-fuel plant.[4]

It's an inspiring idea. But will it survive examination? The first question to ask is whether the technologies work, or, to be more precise, work at something approaching a reasonable price.

The two main sources of renewable power Greenpeace proposes for the energy internet are solar panels and micro wind turbines. I want to believe in them. But the more I read the less convinced I become.

Solar photovoltaic electricity – power produced from panels of light-sensitive cells – is unintrusive and silent. It upsets no one. The infrastructure it requires already exists, in the form of the south-facing roofs on our houses, factories and offices. It is not quite a zero-carbon technology (solar panels in Europe take between two and four years to produce as much energy as is used in their manufacture[5]), but it generates far fewer emissions per watt than our fossil fuel burners. But those who want solar power to supply a high proportion of the electricity used by countries like the United Kingdom encounter two basic problems: there isn't enough sun, and it shines at the wrong time.

The man who has done the most to promote solar electricity in the UK is Jeremy Leggett, chief executive of the company Solar Century, which equips buildings with solar cells. I have known him for years and have great respect for him, not least because, while I sit on my backside telling other people what to do, he spends his time doing it. But some of his statements reinforce my impression that we should be cautious about the claims of those who have something to sell. In his book *Half Gone* – which otherwise has many virtues – he says,

Even in the cloudy UK, more electricity than the nation currently uses could be generated by putting PV roof tiles on all suitable roofs.[6]

This is a big claim to make. Because strong claims require strong support, you would expect it to come from a good source: a peer-reviewed academic journal or a government report, for example. Here is the reference Leggett gives:

'Solar Energy: brilliantly simple', BP pamphlet, available on UK petrol forecourts.[7]

The Energy Technology Support Unit (ETSU), in its report to the government seeking to determine how much renewable energy the country has, calculated that if all the roofs in the United Kingdom were covered in solar panels, and solar electricity could magically be produced at the same rate at all points of the compass, the 'maximum practicable resource' would be 266 terawatt hours (TWh) per year.[8] The UK currently uses some 400 TWh. If my optimistic assumptions about energy efficiency are correct, this could be reduced to 300 TWh by 2030. But unless you are extremely rich, you will not install solar panels at all points of the compass. They are most efficient when they are facing roughly south. If ETSU's estimate were divided by four, that would give us 66.5 TWh. But this takes no account of the cost. When the unit confined its estimates to the amount of electricity which could be produced at 7 pence per kilowatt hour or less (at the time this was roughly the retail price of electricity) it found that the technical potential of roofs in the United Kingdom was 0.5 terawatt hours, or one 800th of our current consumption.[9]

Part of the reason is expressed in the table below.

location	latitude	mean daily solar radiation (kWh/m^2)
Stockholm	59.35 N	2.52
London	51.52 N	2.55
Freiburg (Germany)	48.00 N	3.04
Oviedo (northern Spain)	43.35 N	3.18
Geneva	46.25 N	3.31
Nice	43.65 N	4.03
Porto (Portugal)	41.13 N	4.26
Heraklion (Crete)	35.33 N	4.44
Sde Boker (Israel)	30.90 N	5.70

Source: Tyndall Centre for Climate Change Research.[10]

While solar panels might pay for themselves in Jerusalem or San Diego,* they are a less attractive proposition in London.

The Unit's estimate is far too harsh, however. There are good reasons (which I will discuss in a moment) to assume that the price will fall much faster than it predicted. It might one day make economic sense to clad every south-facing roof, in which case, if ETSU's estimates are correct, solar panels could indeed produce 66.5 TWh – or 22 per cent – of our electricity. But we would immediately encounter the next problem: that the supply of electricity is poorly matched to demand.

In hot countries – as I mentioned in Chapter 6 – electricity use peaks in the middle of a summer day when the air conditioners are turned up. In cold countries, it peaks on winter evenings. If there is one thing of which you can be certain – perhaps the only thing in this field of which you can be certain – it is that the sun does not shine in cold countries on winter evenings.

So even if, by some miraculous technological leap, we were able to produce 300 TWh per year from solar panels on our south-facing roofs, only some of it could be used. There would be a massive surge of production in the summer, during the middle of the day, and hardly anything worth having during the winter, especially when we needed it most. And in a small country like the United Kingdom, when night falls anywhere, it falls everywhere. The energy storage and standby power it required would make a solar economy impossible in a country at this latitude, even if we covered every square metre of ground with panels.

This is not to suggest that solar power is useless here. There is still a high demand for electricity during summer days, and much of that could be met by the sun. If we changed the times of day when we use the most electricity, we might, in the summer, be able to match demand to the surge in production. We could, for example – having loaded them before going out to work in the morning – set our future washing machines and dishwashers to start running at noon.[12] We can also store electricity for some hours in batteries, so the peak

---

* Robert Gross says that 'in sunnier latitudes . . . the costs at 2020 and 2025 become 4p/kWh and 2p/kWh respectively'.[11]

production at midday can be used when we come home in the evening. But it cannot be cheaply stored for use later in the year.

The mismatch between supply and demand also makes a nonsense of another claim propounded by the solar advocates: that a household equipped with solar panels can sell more electricity to the grid than it buys. Technically, this might be true. But it will be selling power when demand – and therefore the price – is fairly low, and buying it when demand and the price are high. Solar panels on our roofs could make a significant contribution – perhaps 5 or 10 per cent – to solving our electricity problem. But to suggest that they could provide the whole answer is unhelpful and misleading.

Here are some estimates of the cost of electricity produced by rooftop solar power:

source	cost per kilowatt hour
Performance and Innovation Unit, 10 Downing Street – present day, United Kingdom	'around 70 pence'
Robert Gross, Imperial College – present day, global[13]	30–80 cents (17–46 pence)
Performance and Innovation Unit, 10 Downing Street – in 2020, United Kingdom	10–16 pence

The table below gives an estimate of how the cost of saving carbon by generating electricity from solar panels in the UK compares to the cost of saving it by other means. The figures are derived from the standard international means of estimating future costs, developed by the International Energy Agency.* This does not make it definitive – nothing in this subject is. The minus figure (for onshore wind) reflects the expectation that by 2020 it will be cheaper than other forms of electricity generation.

Jeremy Leggett correctly points out that the price of small-scale

---

* MARKAL, the IEA says 'was developed in a cooperative multinational project over a period of almost two decades by the Energy Technology Systems Analysis Programme (ETSAP) of the International Energy Agency.'[14]

type of generation	cost in 2020 – low estimate (£ per tonne of carbon saved)	cost in 2020 – high estimate (£ per tonne of carbon saved)
onshore wind	–40	130
nuclear power	105	180
wave	120	430
energy crops	135	185
carbon capture and storage tacked onto existing coal plants	160	200
offshore wind	160	480
new gas plants with carbon capture and storage	180	200
tidal	250	690
new coal plants with carbon capture and storage	460	560
solar photovoltaic energy	2200	3200

Source: UK Department of Trade and Industry.[15]

solar electricity should be compared to the retail rather than the wholesale price of power, because, being on the roof, it is delivered directly to the household without the help of an electricity company. Even so, the estimates in the first table suggest that it is, at present, massively more expensive than either electricity from conventional power stations or electricity from wind farms. (The average retail price was 9.7 pence in November 2005 and 8.7 pence in November 2004.[16]) Even if the price of solar power falls as quickly as the prime minister's office proposes, it could still exceed the retail price of electricity in 2020, not least because the analysts expect that to fall as well. But its price is likely to keep falling, and Downing Street suggests that it 'could become cost competitive with retail electricity in the UK around 2025'.[17]

Jeremy Leggett also makes the point that covering a building with photovoltaic panels can be cheaper than covering it with 'prestige cladding', which is the term architects use to mean fancy façades.[18] This might be true. But solar cladding won't produce much electricity unless it covers south-facing walls not shaded by other buildings. As many buildings with expensive façades are in the commercial districts, where they are likely to be shaded by others, this limits its application.

Here are two facts you seldom see on the same page. Solar photovoltaic cells pay for themselves after 25 to 35 years.[19] Solar photovoltaic cells have a life expectancy of 25 to 30 years.[20,21] At the moment you cannot make your money back. This relationship will soon start to improve, however. Some researchers believe that, with the help of new methods such as the use of silicon spheres, dye-sensitized cells and nanotechnology, the cost of photovoltaic cells could fall as swiftly as costs in the semiconductor industry have fallen.[22,23]

The prospects for micro wind power are less promising. The problem is this: you can produce reasonable amounts of electricity from wind only when it is blowing strongly and fairly consistently. If winds are weak or gusty or turbulent, wind turbines are a waste of time. In built-up areas, winds tend to be weak, gusty and turbulent. Here's what Paul Gipe, the author of one of the leading books on micro wind, says.

Wind turbines should always be located as far away from trees, buildings and other obstructions as practical to minimize the effect of turbulence and maximize exposure to the wind. Turbulence is caused by the wake from buildings and trees in the wind's path. Turbulence can wreak havoc on a wind machine, rapidly shortening its life. Buildings and trees also drastically reduce the energy that is available to a wind machine.[24]

Building for a Future magazine recommends that wind turbines be placed a minimum of 11 metres above any obstacle within 100 metres.[25] So if your house is at least as high as all the houses in your street and the surrounding streets, and as high as any nearby tree, you could get away with an 11-metre pole. If not, you will have to build a minor hazard to aircraft. The higher it is, the less likely you are to receive planning permission, and the more likely you are – because of the lateral thrust exerted – to inflict serious damage to your house. Otherwise, 'A highly turbulent site in which the turbine swings erratically might rob you of 80–90 per cent of your potential energy.'[26] Even if you can find a smooth flow of wind, in built-up areas it won't be strong enough: 'Very few installations are likely to experience more than the equivalent of 4 metres per second average windspeed.'[27]

At an average wind speed of 4 metres per second, a large micro turbine (1.75 metres in diameter is about as big a device as you would wish to attach to your home) will produce something like 5 per cent of the electricity used by an average household.[28] The most likely contribution micro wind will make to our energy problem is to infuriate everyone. It will annoy the people who have been fooled by the claims of some of the companies selling them (that they will supply half or even more of their annual electricity needs). It will enrage the people who discover that their turbines have caused serious structural damage to their homes. It will turn mild-mannered neighbours, suffering from the noise of a yawing and stalling windmill, into axe murderers. If you wished to destroy people's enthusiasm for renewable energy, it is hard to think of a better method.

The only instances in which micro wind turbines might be useful, cost-effective and acceptable to the neighbours is when they have been erected in remote parts of the countryside or built into the structure of tower blocks. In the latter case, they are likely already to be above the line of surrounding buildings. At least a couple of designers have patented aerodynamic towers which will drive wind up into the turbines. One of them, Bill Dunster, claims that his structure raises wind speeds by two or three times.[29,30]

But another source of electricity recommended by Greenpeace is likely to be more reliable. This is something called 'domestic combined heat and power'.

When I report that our power stations have an average efficiency of 38.5 per cent, I mean that the majority of the energy they produce escapes in the form of heat. Using government figures, the energy campaigner Chris Dunham has shown that our demand for heat is roughly equal to the heat wasted by our thermal power stations.*

---

* 'Ignoring non-thermal processes and imports the Department for Trade and Industry gives 978TWh fuel used and 386TWh electricity generated so the average efficiency is 39.5 per cent and a total of 592TWh of waste heat. . . . If we assumed that we could achieve 85 per cent efficiency (45 per cent efficiency for heat and 40 per cent for electricity and 15 per cent wasted), we'd get a figure of 443TWh . . . [The Department] gives 2.39EJ for energy consumption for heat. This equates to 664TWh. But this is energy consumption rather than heat demand which will be lower because of the conversion efficiency. Assuming 80 per cent boiler efficiency this gives 531TWh of heat.'[31]

The micro generation enthusiasts hope to turn two problems into one solution.

The idea, at its simplest, is that instead of using a boiler to produce only hot water and heating for your home, you use a tiny power station to produce heat and electricity at the same time. In 2007, a gas-burning generator called a WhisperGen, which uses an external combustion engine and is already on sale in New Zealand, will be sold in the UK. It costs about 25 per cent more than an ordinary boiler, but the company which markets it says that it will pay for its extra costs within four years, as your energy bill falls by around £150 a year.[32,33] Because both the heat and power are used, the efficiency of an engine like this is somewhere between 70 and 90 per cent.[34] (It produces about twice as much heat as electricity.) It makes no more noise than a refrigerator.[35] In fact, the same kind of engines are used to provide heat and internal power for submarines, because they are so quiet. As boilers tend to die after about twenty years,[36] the nation's entire stock of domestic heating equipment could, with the right incentives, be switched to mini power stations before 2030.

The plan propounded by Greenpeace and others is still more cunning. In the winter, when you want heat as well as electricity, you use your mini power station (a domestic combined heat and power plant) to produce them both at the same time. In the summer, when you want electricity without heat, you use a solar panel. In both cases you can use a battery to store electricity. The Tyndall Centre for Climate Change Research suggests that four industrial lead acid batteries would store enough power to cover the gaps in both summer and winter.[37] Your house, though reliant on gas, would become independent of the grid.

It's a brilliant notion, but will it work? Technically, the answer is 'yes'. The technology is proven, the cost of the heat and power unit (if not the solar panel) is low; and, with the help of the battery, the output of heat and electricity can be closely matched to people's demands. The WhisperGen's external combustion engine* produces about 6.6 times as much heat as electricity.[38] Friends of the Earth calculates that if 10 million homes installed heat and power units,

---

* The famous Stirling engine.

they would produce about 30 terawatt hours of electricity a year – 7.5 per cent of the country's total requirement.[39] There are 24.5 million households in the United Kingdom,[40] and they consume some 29 per cent of the UK's electricity,[41] so if everyone equipped their homes with this machine, we'd generate about 74 per cent of our own electrical power. Other machines have a better ratio of electricity to heat: in principle, our homes could produce all the electricity they need from combined heat and power burners.

Unfortunately, the contribution these mini power stations make to preventing climate change is less convincing. The company which markets the WhisperGen system claims that it will save about a fifth of the carbon a household's supply of gas and electricity would otherwise produce.[42] A report by the Energy Saving Trust, which is funded by the government, estimates that if micro generation (a mixture of combined heat and power, solar electricity and micro wind) supplied our homes with 220 terawatt hours of energy – which is about 42 per cent of the energy our houses use* – it would save just over 6 per cent of our household emissions.[44] This suggests that if we generated all the energy we needed in our homes by micro generation, we would cut our household carbon emissions by just 14 per cent. And the report is more generous to micro wind than it deserves.

District heating systems use the same principle – combined heat and power – but on a slightly larger scale. Heat from a small power station, serving a tower block, a housing estate, a suburb or perhaps even a whole town, is circulated by pipes into people's homes. This system, like all good things, is widely used in Scandinavia. In Helsinki, for example, 98 per cent of heating comes from district schemes of one kind or another.[45] The advantage of district heating, especially if it serves offices and factories as well as homes, is that the power supply doesn't have to fluctuate as much, as demand is more consistent.[46] It can make more effective use of other sources of heat, such as the ground-sourced heat pumps I discussed in Chapter 6. It is also popular in Germany, supplying, for example, the Reichstag in Berlin. The

---

* The domestic heating demand in the UK, according to a report from AEA Technology, is 452TWh per year [43], while lights and appliances in our homes, according to the Environmental Change Unit, use 73TWh of electricity. The total energy demand is 525TWh.

surplus heat the Reichstag's system produces in the summer is pumped, in the form of hot water, into porous rocks 300 metres below the surface, then pumped back up to heat the building in the winter.[47]

Regrettably, district heating systems are harder and more expensive to attach to an existing development than to incorporate into a new one: streets, pavements and floors have to be excavated to lay the pipes.[48] But it is difficult to think of a good reason why district heating should not be installed whenever a new housing estate is built. On second thoughts, there is one good reason: that the new homes would be so well constructed they need no heating at all.

If micro generation is to become a more effective means of reducing carbon emissions, it must switch from gas to another kind of fuel. I have already explored the potential, with a growing sense of hopelessness, of biomass, biogas, solar hot water, ground-sourced heat pumps and geothermal energy. Some of these could be used to help decarbonize the energy internet – wood chips or pellets, for example, work well in district heating systems – but their potential contribution, as I have shown, is limited.

This appears to leave us with just one option: the most abundant element on earth. Hydrogen behaves rather like a fossil fuel, in that it burns with a nimble flame and can provide several different kinds of power. But it produces, when it combusts, nothing but water. So we should not be surprised to find that it has become the environmentalists' Red Lion or Powder of Projection: the alchemical catalyst which could change everything.

There are several means of producing it. Unfortunately their costs, as costs often are, are inversely proportional to their environmental impacts. The table below gives the figures produced by the National Academy of Engineering in the United States. These prices include distribution and delivery.

Hydrogen is made from coal mostly by bashing the fuel into powder and passing steam and oxygen through it. It is made from natural gas by heating it and reacting it with steam.[49] It is made through electrolysis – as anyone who didn't manage to escape from their chemistry lessons at school will remember – by passing an electric current through water. The last method is the philosopher's stone. The electricity could be produced from renewable power. You could generate

method of production	cost, today (US$ per kg)	cost, future (US$ per kg)
from coal	1.91	1.41
from coal with carbon capture and storage	1.99	1.45
from gas	1.99	1.62
from gas with carbon capture and storage	2.17	1.72
from electrolysis	6.58	3.93

Source: The National Academy of Engineering.[50]

from thin air and water a fuel which is dense, storable and transportable. If it could be produced, carried and burnt cheaply, it could replace our fossil fuel economy while ensuring that we need scarcely change the way we live.

But it shouldn't be hard to see that the last method will always be expensive. Electricity is a refined manufactured product. Using it as a raw material to produce another kind of fuel is a novel species of extravagance. The National Academy says that the efficiency of producing hydrogen from electricity is just 30 per cent.[51] In other words, 70 per cent of the useful energy produced by our wind turbines or power stations would be lost. The re-forming of natural gas, by contrast, is about 72 per cent efficient.[52]

If renewable electricity is priced at just 2 pence per kilowatt hour, hydrogen produced by this means and delivered to people's homes would cost twice as much as natural gas.[53] Hydrogen produced from natural gas, even after the carbon dioxide has been captured and stored, is likely to cost households only 50 per cent more than the gas itself.*

---

* The Imperial College Centre for Energy Policy and Technology gives the prospective delivered cost of hydrogen to residential customers as 4.8 pence per kilowatt hour, when the hydrogen is produced by electrolysis.[54] Of this, 2.5 pence is incurred by hydrogen production: the rest is accounted for by compression, storage and distribution. The NAE's figures suggest that in 'the future', hydrogen produced from gas with capture and storage will cost 44 per cent as much as hydrogen produced by electrolysis, which means that production, if these figures hold, will cost 1.1 pence per kilowatt hour, bringing the total cost of hydrogen delivered to residential customers down to 3.4 pence. In November 2005, natural gas cost households 2.3 pence/kWh. In November 2004 it was 2.0 pence/kWh.[55]

Because, for the reasons I discuss below, we would need to consume less hydrogen than natural gas to produce the same amount of electricity and heat, this means the price we pay for fuel would stay roughly the same. But because the device in which it is burnt, at least in the early years, is likely to cost more, as I will show below, the overall price of heating and electricity will be higher than it is today.

So at first sight it seems that the re-forming of gas, accompanied by carbon capture and storage, should supply our hydrogen. Where there are power stations today, there should be hydrogen factories tomorrow. Eighty per cent or more of the carbon dioxide they produce can be buried.

In the carbon capture and storage system I described in Chapter 5, the carbon dioxide is stripped out of the exhaust gases produced when natural gas is burnt to make electricity. In the case of hydrogen, it is separated and captured before the gas is burnt. Apart from that, there is no substantial difference: the carbon dioxide is piped away and buried as before.

At first, hydrogen could be mixed with natural gas, using the existing pipeline network: a concentration of 10 or 15 per cent would apparently make no difference to the performance of our boilers.[56] But as we began to produce more of it, and to install machines designed to burn it, it would have to be transported separately. Liquefying the gas (because it has to be kept at minus 259° or below) uses up about 35 per cent of the energy it contains.[57] Piping it by itself would require an entirely new network, as pure hydrogen would cause our gas pipes to become brittle.[58] We could, of course, use the existing pipelines to shift natural gas very close to where the hydrogen will be needed and build small steam re-forming plants to produce it on the spot. But then the carbon dioxide could not be buried. So it seems as if the only viable options are either to build a new piping system, alongside the gas pipes, to move molecular hydrogen around the country; or to replace the natural gas pipes with hydrogen pipes as they reach the end of their lives; or to shift it by truck, ship or train. All these means appear to be viable, though pipelines are likely to be both more energy-efficient and more convenient for the customers. The US National Academy of Engineering reports that, while molecular hydrogen 'is a uniquely difficult commodity to ship on a wide scale',

about 9 million tons of hydrogen are manufactured annually in the United States and transported for chemical and fuel manufacturing as a low- or high-pressure gas via pipelines and trucks . . . Much experience worldwide has been achieved over many years to make these transportation modes safe and efficient.[59]

The energy needed to compress it to 5,000 pounds per square inch (which is two and a half times higher than required for high-pressure piping[60]) amounts to between 4 and 8 per cent of the energy the hydrogen contains.[61] Transporting it as a gas at high pressure, in other words, is much more efficient than transporting it as a liquid. Building a hydrogen pipeline is expensive, however. The pipes must be 50 per cent greater in diameter than gas pipes, and the materials of higher quality.[62] These costs have already been taken into account in the calculations I made above, which suggest that hydrogen delivered to consumers will cost about 50 per cent more than natural gas.

There might be another way of doing it, however. As the energy expert Dave Andrews suggested to me, if hydrogen were produced from electricity by means of mini-electrolysers in the home, the waste heat (which accounts for the 70 per cent loss of energy when extracting hydrogen from water) could be used to keep the house warm. This would help to pay for the costs of conversion. When the wind was blowing strongly, the surplus power it produced could be used to generate hydrogen and heat. When it dropped, the hydrogen could be used to provide heat and electricity. This means, however, that hydrogen would have to be stored at home.

In either case, hydrogen could be burnt in boilers very similar to those which use natural gas: the difference would be that it produced no carbon dioxide emissions. But a further possible advantage of using this element is that it can also produce both heat and electricity in a device called a fuel cell. This is a kind of battery, which uses a gas as its chemical energy source rather than a solid or a liquid. It generates a higher proportion of electricity to heat than an external combustion engine such as a WhisperGen, and no effluent but water.

It makes no noise and responds almost instantaneously to demand. Already, fuel cells exist with electricity production efficiencies of around 60 per cent.[63] This is probably too good for our homes: a less-efficient fuel cell would match the ratio of heat to electricity we need more closely.

Fuel cells are currently more expensive than other kinds of micro generators: even the most optimistic estimate suggests that if they were widely commercialized today, they would be around four times the price, for example, of a diesel engine with the same output.[64] Most authorities suggest they are more expensive than this. But with an installation programme backed by the government, the cost could fall swiftly. The Tyndall Centre for Climate Change Research says that combined heat and power systems run by fuel cells may be 'economically viable by 2009'.[65] They will also need to be scaled down if they are to be widely used in houses: at the moment they are several times as big as combined heat and power units running on natural gas,[66] which are about the size of an ordinary boiler. Their lifetime, which is currently quite short, needs to be extended. Taking all this into account, it might be more realistic to consider replacing our gas boilers with hydrogen boilers providing heat and electricity, rather than with fuel cells.

But both boilers and fuel cells are useless without fuel. Micro generation of this kind, unless it uses domestic electrolysers, is impossible without the creation of a hydrogen storage and delivery network. As a hydrogen storage and delivery network would be useless without micro generators to supply, we are confronted with the classic chicken and egg problem of a switch to a new technology. Neither half of the system can function until the other half does. In other words, the establishment of a hydrogen distribution system would need to be assisted – both strategically and financially – by government.

It would be foolish to pretend that there are no environmental problems associated with a hydrogen economy. If hydrogen leaks and finds its way into the troposphere, it becomes, by indirect means, a greenhouse gas.* In the stratosphere, it appears to accelerate the

---

* It reacts with hydroxyl molecules to form water vapour. Hydroxyl currently mops up some of the methane in the atmosphere[67]

depletion of ozone.[68] Any system we use must be well sealed. If the hydrogen is made from coal, as the US government proposes,[69] this will result in a massive boost for the quarrying industry. Even if it is made from natural gas, it means sustaining our dependence on a fossil fuel, which is often associated with ugly infrastructure, dodgy deals with unpleasant governments, habitat destruction and the displacement of local people.[70]

But the gas does appear to provide us with a fairly cheap means of generating heat and electricity, while greatly reducing our carbon emissions. Taking into account the losses involved in carbon capture and storage and hydrogen transport, we could achieve a total carbon saving of about 80 per cent. The combination of the energy efficiencies I discussed in Chapter 4 and new kinds of fuel pushes us beyond my target of a 90 per cent cut in household carbon emissions. But to get there would require a massive and extremely ambitious government programme, swapping the natural gas distribution system for a hydrogen network. It also means that every boiler which dies between 2010 and 2030 must be replaced with a boiler burning hydrogen or a hydrogen fuel cell. This is a tall order, but it could – just – lie within the realms of possibility. Household electrolysers, producing hydrogen from low-carbon electricity, could be quicker and easier to develop and install than a new pipeline network. The cost of the fuel is likely to remain higher, however, though it would be suppressed if the electrolysers work only when there is surplus wind power.

But if we did scrap the national grid, could a micro power system work, without repeatedly plunging us into darkness? The answer, which I realize is not entirely reassuring, is 'probably'.

Every household is linked to its neighbours to form a miniature version of the national grid: a generating 'island'. This island can in turn be connected to the surrounding micro grids to offer more security. Electricity can be automatically traded, with the help of smart meters, between households: if someone's generator fails, others can fill the gap.[71] There will be little need for extra backup generators, as everyone's machine supports everyone else's.

Greenpeace says that this system will be more secure than the national grid:

It would deliver an electricity supply far less vulnerable to massive system failure as a result of sabotage or extreme weather.[72]

The House of Lords suggests it would offer the same degree of security:

This model is in principle no better or worse than the national one for responding to coherent changes in demand . . . There would be technical issues to resolve, including the need for synchronization between islands, but these could be overcome.[73]

The Tyndall Centre starts confidently:

We find that there is no fundamental technological reason why microgrids cannot contribute an appreciable part of the UK energy demand . . .

but later qualifies its enthusiasm:

In a microgrid, frequency stability becomes critical . . . because the system is small the problem is much more difficult to manage to the same standard as is normal in a utility system . . . Control of power quality will be the biggest issue for a microgrid. Voltage dips, flickers, interruptions, harmonics, DC levels, etc., will all be more critical in a small system with few generators.[74]

But these problems, it says, can be overcome with the help of batteries, which will respond 'fast enough to ensure adequate frequency control'.

Greenpeace argues that the costs of a system like this would not be as high as they first appear because we would not have to pay so much to maintain and rebuild the national grid, or pay for the electricity lost in transmission (about 7.5 per cent of the total[75]). It points out that only 52 per cent of the capital cost of electricity is incurred by generation: the rest pays for carriage and distribution.[76] The International Energy Agency estimates that electricity suppliers in the European Union must invest $1.35 trillion between now and 2030 to keep the system running, of which $648 billion must be spent on the transmission and distribution networks.[77] If some of this money were instead spent on building hydrogen pipelines, the switch to an energy internet would start to look less painful.

But, exciting as the idea is, I have come to believe that the dismantling of the national grid would be a mistake: it would be to succumb to the aesthetic fallacy. There is still a huge demand – from industry, offices and the people who light our streets and run our trains – for electricity which isn't accompanied by heat. In fact, it seems to me that far from shutting the network down, we might need to expand it. New cables would enable us to extract electricity from the best ambient sources, far from our coasts. By drawing power from a large area of land and sea, a wider grid would make renewable resources more reliable. By connecting us to the networks of other countries, and the wide range of electricity sources they can exploit, it could reduce the 'system costs' of keeping demand and supply in balance. I hate pylon lines, but I cannot help concluding that we need more of them.

So I might have the answer. Our homes have the greatest requirement for heat. A micro-generation system using solar panels and either hydrogen boilers or hydrogen fuel cells would supply their heat and their electricity. Either they could make their own hydrogen from electricity supplied by the grid, or they could obtain it from a pipeline network. Perhaps most importantly – as far as public acceptability is concerned – this system demands no more time and trouble for householders than they currently spend on managing their supplies of energy: in other words, none whatsoever. Everything comes on and goes off at the flick of a switch and, in principle, works as smoothly as our heat and electricity systems do today.

Around half of our grid-based electricity could be supplied, as I suggested in Chapter 6, by means of a few very large power stations burning methane – either in the form of natural gas or the effluvium from underground coal gasification – and burying the carbon dioxide they produce. The other half, if my meta-guess is correct, could be provided by offshore wind and wave machines.

This transformation would require some bold politics and some ambitious engineering. But it isn't a manned mission to Mars. Everything I have proposed here is, as far as I can tell, already technically possible and more or less economically feasible. The question is whether it can be done in time. If it can, I might have saved myself from frightful nemesis. I will not have to become an aromatherapist.

8

A New Transport System

Are any Britons here? They're always travelling.
Faust, Part II, Act II[1]

In 1929, long before manmade climate change had been detected, the Russian journalist Ilya Ehrenburg wrote this.

the automobile . . . can't be blamed for anything. Its conscience is as clear as Monsieur Citroen's conscience. It only fulfils its destiny: it is destined to wipe out the world.[2]

He had seen something the rest of us are only beginning to comprehend.

The production of carbon dioxide by land transport should be easier to solve than the other problems this book addresses. Everyone can see how inefficient the transport system is: thousands of people, one to each overpowered car, heading in the same direction every day then heading back again. The technologies and economic policies required to address it have been available for decades. Far from costing more money, a rational, efficient system, producing 10 per cent of current emissions or less, would save us billions. But the real problem is neither technological nor economic. It is political or, more precisely, psychological.

In managing our transport systems, our governments must constantly negotiate the paradox of mass movement. They must create a system which, for the sake of speed and efficiency, treats us like a herd, constantly prodded and coralled, divided, re-formed and forced into line. At the same time it must grant us the illusion of autonomy. The songs of the open road, the pictures in the advertisements of cars on lonely mountain passes, the names of the vehicles, especially the all-terrain vehicles which seldom venture beyond the suburbs – *Defender, Explorer, Pathfinder, Cherokee, Touareg* – all speak of a freedom which is not to be found on our highways. But this is the fantasy that governments must strive to preserve.

Having freed ourselves, with the help of fossil fuels and the technologies they permit, from the constraints of the past, we have embarked not upon the wild life of the spirit but upon a life of compression and self-control. Thanks to the blessings of modern medicine and agriculture, we live in crowded countries, in which the freedom to swing your fist leads invariably to someone else's broken nose. Prosperity raises fences: as we become richer we can travel further, but there are fewer places into which we can venture.

Restrained by our freedoms, polite, obedient, safe, we lead lives of metaphor – metaphor that constantly invokes an age of ecological and social hazard. In the financial press the abstractions of money are illustrated by bulls, bears and tigers, sharks and mammoths, men lost in woods and wolves in sheep's clothing. Until 2005, the highest-paid executive in the world, Michael Eisner of Disney, ran a corporation whose core business was that of investing animals with human characteristics, a practice by which the first hunters, living in ecological time, came to know their prey.

In 1990 I stayed for a few weeks in a world released from sublimation. In the *garimpos* – the informal goldmines – of northern Brazil, in the rainforest hundreds of miles from the nearest town, lived an ephemeral society untroubled by the law. Disputes were resolved by gunfire: within six months, 1,700 of the 40,000 miners there were killed.[3] Wealth was not earned but dug from the ground.

The mines represented everything I have spent my adult life fighting. Governed by violence, occasionally all-out war, they tore open one of the richest ecosystems on earth, and introduced to the Yanomami Indians guns, disease, alcohol and prostitution. But nothing there shocked me as much as myself. I loved it. In the *garimpos* I found the freedom I had always been promised but had never seen. I didn't use it. I killed no one, felled no trees, dug no gold. But I could have. The mines were the metaphor. They were the inner life I had led, of which, until then, I had scarcely been aware.

The freedom of the road, of that strip of concrete laid down in order to herd us better, is an expression of the tension between civilization and the life of the spirit. Under the bonnet lurks a beast which could, were it not hemmed in by social and physical restraint, roar into murderous speed – 120, 140 miles per hour. The opportunity the engine forgoes is another manifestation of the metaphor: the inner life of both the car and the mind. The faster cars become and the more the roads fill with traffic, the greater the tension between the metaphor and reality.

I believe that the growth in driving is one of the primary reasons for the libertarianism now sweeping through parts of the rich world. When you drive, society becomes an obstacle. Pedestrians, bicycles, traffic calming and speed limits become a nuisance to be wished away. The more you drive, the more you seek the freedom that the road promises but always denies. From time to time this frustration takes political shape, as motorists blow up speed cameras[4] or, when petrol prices rise, blockade the oil refineries. But the new libertarianism is also now driving down the pavement. Organizations such as the Association of British Drivers, which began by campaigning against speed cameras and road humps, have now joined forces with people calling for lower taxes and the destruction of the 'nanny state'.[5] The car has become an agent of political change.

Governments, while pretending to do otherwise, indulge the fantasy of freedom as much as they can. Whenever they introduce the slightest restraint upon the available road space or the behaviour of the people who use it, the tabloid newspapers start thundering about 'the war against the motorist'. Our governments know – as a British transport minister first admitted in 1938[6] – that demand is a function of capacity. The more roads you build, the more the volume of traffic will rise to fill them. Unless they wish to cover the entire country with concrete, they know that they cannot keep accommodating rising demand. So from time to time they announce policies which are supposed to dissuade people from using their cars. As John Prescott, the Deputy Prime Minister of the United Kingdom, promised in a rare moment of lucidity in 1997:

I will have failed in this if in five years there are not many more people using public transport and far fewer journeys by car. It is a tall order but I want you to hold me to it.[7]

Between 1997 and 2004, car journeys in the United Kingdom increased by 9 per cent.[8] The government expects this growth to accelerate:

The central projection is for traffic to grow by 26 per cent between 2000 and 2010, implying an annual average increase of 2.3 per cent over the whole decade (around 2.6 per cent per year for the rest of the decade).[9]

One of the reasons for this can be found in the paper it published in 2005. The graphs there show that though gross domestic product has risen by 60 per cent since 1986, the cost of driving has fallen.[10] The government expects the two lines to keep diverging to 2025 and beyond. While bus and coach fares have risen in real terms by 66 per cent since 1975, and train tickets by 70 per cent, the cost of owning and running a car has fallen by 11 per cent.[11] This gulf seems to have widened since Mr Prescott took office. Rail and bus fares have risen by 7 per cent and 16 per cent respectively since 1997, and motoring costs have declined by 6 per cent.[12]

To give him his due, half of Prescott's promise has been met. The number of people using public transport has also been rising:

Our central projection shows rail passenger kilometres increasing by 33 per cent between 2000 and 2010. This implies a pick up in growth to around 3.5 per cent a year for the rest of the decade ... Growth in bus patronage over 2000 to 2010 is projected at around 11 per cent.[13]

This, as far as climate change is concerned, is the worst of all possible worlds. Instead of replacing cars, trains and buses are supplementing them. Transport which does not require the use of fossil fuels, by contrast, has declined. In the past ten years, according to the National Travel Survey, the number of walking trips has fallen by 20 per cent.[14]

In 2004 we cycled 6 per cent less than we did in 1992.[15] Many of those who once walked or cycled now use cars for the same trips instead: one quarter of all car journeys cover less than 2 miles.[16] The table below shows how the United Kingdom's journeys are distributed. The government does not include walking and heavy goods transport in these figures.

mode of transport	annual distance travelled (billion passenger kilometres)	percentage of total
cars, vans and taxis	679	85.2
trains	51	6.4
buses and coaches	48	6.0
planes[i]	9.8	1.2
motor cycles	6	0.8
pedal cycles	4	0.5
Total	**797**	

i. The government counts only domestic journeys. A more accurate figure would include 50 per cent of the length of international trips.
Source: Department for Transport.[17]

Excluding international air travel and shipping, transport accounts for about 22 per cent of our carbon emissions; 91 per cent of these are produced on the road. The government expects the carbon produced by road transport to rise by between 5 and 7 per cent this decade.[18] It proposes, in effect, to do nothing to prevent it.

The main policy for reducing carbon emissions from road transport is the voluntary agreement with European, Japanese and Korean car manufacturers, which aims to reduce carbon emissions per kilometre from new cars by 25 per cent on 1995 levels by 2008.[19]

This policy (as the government's projections anticipate) has already failed. In July 2005, *Automotive News* reported this.

'Carmakers are very far from reaching their 2008 target,' said Patrick Coroller, an official at France's Agence de l'Environnement et de la Maîtrise de l'Énergie (ADEME), a state-funded environmental agency . . . Japanese and Korean carmakers are even further away.[20]

While invoking miracles which it already knows will not materialize, the British government continues to promote the growth of traffic by building more roads: as I mentioned in Chapter 3, it is spending £11.4 billion on creating more space for traffic,[21] knowing that this space will inevitably be filled.

Though journeys by buses and trains can do nothing to reduce carbon emissions while they supplement travel by car, if all the country's drivers were magically whisked into public transport, a 90 per cent cut could almost be delivered in one stroke. The table below gives the government's figures for the carbon dioxide emissions produced in travelling from London to Manchester. It assumes that the car is of average size and contains the national mean of 1.56 people, that the train is a modern electric model and 70 per cent of its seats are occupied, and that the coach has forty passengers.

mode of transport	carbon dioxide emissions per passenger (kg)
car	36.6
train	5.2
coach	4.3

Source: written parliamentary answer.[22]

Two conclusions can be drawn from these figures. The first is that train travel, which in the public mind is equated with low environmental impacts, is less efficient than travel by coach. If we were to put all considerations but climate change to one side, the railways would best be cleared of human traffic and used instead to take long-distance freight off the roads. According to the Strategic Rail Authority, transporting goods by lorry produces 180 grams of carbon dioxide per tonne per kilometre, while transporting them by train emits 15: a saving of 92 per cent.[23] At the moment, just 12 per cent of our freight travels by rail.[24] But the figures in this table are not fixed. There is plenty of scope for improving the carbon accounts of the railways: Japanese trains, for example, are much lighter then European ones.

The second is that if you switch from car to coach, you cut the carbon you would otherwise have produced by 88 per cent. With some very small efficiency improvements, universal coach travel, if the total mileage we covered did not increase, would meet my target.

The trouble is that people hate coaches, and for good reason. Coach travel is a dismal and humiliating experience. When I take the bus, as I sometimes must, from Oxford to Cambridge, I arrive feeling almost suicidal. First I must cycle for 20 minutes in the wrong direction, into the city centre. Then, usually in horizontal sleet and clouds of diesel fumes, I must wait for a man who looks as if he has just drunk a quart of vinegar to grunt that the bus is ready for boarding. I give my money to someone who makes the other man look cheerful and sit on a chair designed to extract confessions. Then, weaving around bicycles and bollards, the coach fights its way through streets designed for ponies. After half an hour it leaves the city. It then charts a course through what appears to be every depressing dormitory town in south-east England, hoping to pick up more custom. On a good day, with a following wind, the journey from my house to my destination in Cambridge, a total of 83 miles, takes four and half hours. The average speed is 18 miles an hour, about 50 per cent faster than I travel by bicycle. If I made the journey by car, I could do it in 100 minutes.

Unsurprisingly, coaches, on most routes, have become the preserve of those who can't afford to travel by other means. In both the United Kingdom and the United States they are associated with poverty and exclusion, with people who have time to spare, but little money. Quite aside from the fact that it offers us no fantasy of freedom, a switch from private to public road transport is counter-aspirational, and therefore of little interest either to wealthy travellers or to our political leaders.

It would work politically only if coach travel could, in some way, be presented as a better thing than private transport. It seems to me that there is just one means by which this could be done: to ensure that it is swifter, more relaxing and more reliable than the motor car.

Because everything possible has been done to smooth the passage of the car, this sounds, to say the least, challenging. But an unpublished paper by the economist Alan Storkey appears to have cracked it.[25]

Storkey's key innovation is to move the coach stations out of the city centres and on to the junctions of the motorways. One of the main reasons why long coach journeys are so slow in the UK (often averaging 20 miles per hour or less) is that in order to create a 'system' – which allows passengers to transfer from one coach to another – they must enter the major towns, travel all the way to the centre and all the way out again. In the rush hour, you might as well walk. A car travelling from one end of the country to the other, by contrast, can stick to the motorways throughout its journey. Our city centre stations, as Storkey points out, are historical accidents: the coaches they were built to serve were pulled by horses. They were probably almost as fast.

Instead of dragging motorway transport into the cities, Storkey's system drags city transport out to the motorways. Urban buses, he remarks, currently serve the distant suburbs very badly. If the buses (or for that matter, trams or light railways or underground lines) travelled a little further, to the nearest motorway junction, they could take people from their homes or to their destinations without affecting the speed of the coaches. If they used dedicated bus lanes, and the buses were given priority at junctions, they could move people into and out of the cities at busy times faster than travel by car. By connecting urban public transport to the national network, Storkey's proposal could revitalize both systems, as it would greatly increase the number of passengers while providing better services for the suburbs on the way to the motorway junctions.

Unlike train companies, which are economically viable only when their locomotives can pick up dozens or hundreds of people at every station (which means they must travel infrequently), coach firms can afford to deploy much more stock.

With 200 coaches on the M25 there would be a coach within a mile of a stop most of the time in both directions, and the waiting time would be some two or three minutes ... With an effective network many of the inter-city motorway journeys could have a waiting time of five minutes or less. Once people could use this system with confidence in its regularity, demand would surge. Then high-occupancy coaches could emerge.[26]

To enhance their speed and regularity, the coaches would be given dedicated lanes on the motorways and the orbital roads. Traffic lights responding to a radio frequency would grant them priority at junctions and roundabouts (ambulances in this country already have transponders that turn the lights green). The tabloid newspapers might fulminate, but it would not be long before people stuck in their cars began to notice the buses roaring past on the inside lanes.

There is a sound mathematical reason why people in coaches travelling in dedicated lanes will move faster than people on the rest of the road. The safe stopping speed of a car travelling at 30 miles per hour is 23 metres, and at 60 miles per hour, 73.*[27] Storkey uses the M25 (the vast orbital road around London) to show how this affects motorway capacity. The road is 118 miles long. It has three lanes in each direction most of the way, but four lanes along 35 per cent of its length. The table below shows the numbers of people it can accommodate.

average speed (miles per hour)	number of cars	number of passengers[i]
30	33,050	52,880
50	15,655	25,048
60	11,589	18,542
70	8,924	14,278

i. Assuming an average occupancy of 1.6.
Source: Alan Storkey.[28]

I had to rub my eyes when I first read these figures. When the cars are travelling at 60 miles an hour, the entire motorway – all 6.7 lanes, 790 miles in total – can accommodate just 19,000 people. Every coach, by contrast,

hoovers up a mile of car lane traffic . . . This mode of transport can carry nearly five hundred people in a mile of roadway, not a mere thirty, and the capacity of the M25 moves from 15,000–20,000 in cars to 260,000 in coaches.[29]

----

* It doubles in wet weather.

Released from congestion, the coaches are less likely to bunch up. Some services would constantly circle the orbital roads around our major cities. Others would travel up and down the motorways that link them.

In order to achieve average speeds of 50 mph or more, coaches need to travel at 65 mph and have stops at intervals of ten or twenty miles. This means that stops do not necessarily occur at every motorway junction.[30]

When the faster urban links Storkey proposes and relief from the need to find a parking space are taken into account, this could bring the overall average journey time to below that of car travel. At rush hours and on bank holiday weekends the public system could be very much faster. While drivers would lose their sense of autonomy and the convenience of being able to sit, without shifting themselves or their luggage, in one place throughout the journey, they would gain time to read, watch films, make phone calls or sleep.

To make this work, the coaches will need

good leg room, seat quality, work stations, food, drinks and media stations … In other words these vehicles can be an elite form of road travel … Coaches, in principle, are another form of stretch-limousine.[31]

This might mean losing a certain amount of efficiency, as they couldn't accommodate quite as many people as they do today, but this would be offset by the fact that the coaches no longer need to travel in and out of the city centres, which reduces the distance they must travel. Automated luggage systems would make the stops much shorter. In other words, the country's slowest, most uncomfortable and most depressing form of mass transport could be transformed into one of its fastest, smoothest and most convenient systems. Instead of feeling like social outcasts, its passengers – watching films and making calls while they flit past the cars stuck in their congested lanes – could begin to see themselves as the kings and queens of the road.

In terms of what we usually spend on transport, Alan Storkey's system costs next to nothing. It requires no new roads, no railway lines, no significant subsidies. Coach stations need to be built, new lines need to be painted on the motorways, coaches bought and the traffic signalling system adjusted. Storkey suggests his proposal would cost about £1 billion to implement. This is 27 per cent of the price of widening the M1. But even this, I think, is an overestimate, for he does not take into account the money to be made by selling the land in the city centres that is currently used for coach stations. It seems to me that his system is likely to be self-financing from inception. And it releases some of the staggering sums of government money used to keep the existing system alive.

The motorists' lobby points out that drivers pay about £30 billion a year in tax, compares this figure to the cost of road building, and claims that drivers are subsidizing the government. But as Storkey points out, the government also pays for policing and ambulances and the health costs of people affected by local pollution* or caught in traffic accidents. If roads, like railways, were to generate an 8 per cent return on their asset values, the income forgone by the government would amount to £32 billion a year.[35] As coaches are very much safer than cars† and (per passenger) produce less pollution, both the costs which can be quantified and those which cannot are likely, under his system, to fall. The private costs of transport will decline more dramatically.

> Capital investment in coaches is far more productive . . . than our present use of cars, because they are used so much more intensively. One piece of equipment costing, say, £150,000 can carry people on nearly 100,000 substantial person/journeys a year, while a car costing £15,000 would be likely not to clock up 1,000 such journeys. This is a tenfold increase in the efficiency of capital use.[36]

According to the Royal Academy of Engineering,

> Congestion already costs the UK an estimated £15 billion each year and

---

* According to the European Union, 40,800 deaths a year in the United Kingdom are caused or hastened by air pollution, most of which comes from vehicles.[32] It gives an estimate for the total cost of air pollution per person per year in the UK of between €1,341 and €2,497,[33] which would mean a total cost to the UK of €81–151 billion (the population in July 2006 was 60.6 million[34]). But, as I explained in Chapter 3, I am very suspicious of these life-costings.

† Storkey notes that 'coaches had a zero level for fatalities per billion passenger kilometres throughout the decade 88–97 and a level of serious injuries at one third the level of cars.'

that figure is expected to double over the next decade as more of our transport system reaches capacity.[37]

This is not the only cost that will keep rising. If it is true that global oil production will peak soon then go into decline, profligate fuel use will become not only unaffordable but also impossible: all current transport systems will more or less grind to a halt. The UK government's plan for addressing congestion – a road pricing system reliant on perpetual surveillance and satellite monitoring – will be extremely expensive, punitive for the poor, far more intrusive and coercive than Storkey's system and of little use in reducing carbon emissions. Again, even if climate change were not happening, a transformation of the kind Storkey describes would be necessary.

But when a proposal similar to his, though on a smaller scale, was put forward by the government's consultants (they proposed a coach system around the M25), the government crushed it. The transport secretary, Alistair Darling, was asked in the House of Commons 'why has he specifically rejected the recommendation ... for a strategic authority to create a high quality orbital coach network?'[38] This was his response:

Given everything that the honourable gentleman said about bureaucracy, I am astonished that his one new policy announcement is that he wants a strategic authority for coaches. I should have thought that running buses and coaches was best left to existing organizations.[39]

It's reassuring, isn't it, to see how seriously the government takes these questions.

Storkey's proposal does not solve all our transport problems. As a means of addressing climate change, it would work only with the help of two other policies: the capping and rationing of the carbon we use, and the capping and rationing of the road space we use. Otherwise it merely liberates space into which traffic can expand, while releasing money for investment in energy-intensive processes, as the Khazzoom–Brookes Postulate, which I explained in Chapter 4, predicts. Reaping the carbon benefits of Storkey's proposal means giving buses and coaches some of the lanes that cars now use, rather than building new lanes to accommodate them. In city centres, it means reclaiming some of the streets now used by cars for pedestrians and cyclists, as Copenhagen, Vienna and Zurich have done. It means turning car parks back into city squares, and planting trees and installing playgrounds and pavement cafés where there was only tarmac before. It is important, too, that Storkey's coach stations on the motorway junctions do not become new development hubs: it is easy to see how governments could use them as an excuse to grant planning permission for new superstores, service stations and housing, 'integrated with public transport'. Developments like this could counteract many of the scheme's savings.

Reducing the number of cars on the roads is not just a matter of providing alternatives, but also of discouraging driving. A carbon rationing system, like the congestion charge which has reduced the traffic in London, provides a powerful financial incentive to switch to public transport. With these additional policies, Storkey's coach system provides the means by which people can stay within the necessary carbon limits while travelling almost as much as they do today. While the fantasy of freedom would have to be abandoned, real freedoms would be preserved.

There are some people who would still need cars. Trains and buses are almost impossible for some disabled people to use, and it is hard to see how a public transport service could reach everyone who lives in the countryside without either costing a fortune or restricting their current freedoms. People who live in cities will find that there are some journeys they cannot easily make by coach, though Storkey's system should ensure that it is more cost-effective to hire a car when you need it, or to join a car club, than to own one. If some cars are to remain on the roads, sustaining a 90 per cent cut in carbon emissions means both improving their efficiency and reducing the need to travel.

Confronted with the twin disasters of climate change and an impending oil peak, it is hard to see how anyone could justify the assertion that the need to drive a car which can accelerate from 0 to 60 miles an hour in 4.5 seconds (the Audi S4 for example) overrides the Ethiopians' need to avoid recurrent famines, or the whole world's need to avoid the economic catastrophe we'll suffer if petroleum peaks too soon.[40] The speed and acceleration of our cars is a form of profligacy at which all future generations will goggle. The main reason why improvements in energy efficiency have been so slow to take effect is that manufacturers insist on sustaining performance. If this were not a requirement, the most efficient engines would be many times smaller than they are today.

The average emission for new cars in the United Kingdom is 170 grams of carbon dioxide per kilometre.[41] The Toyota Prius, rated by the US Environmental Protection Agency as the greenest car on the mass market, manages 104 grams,[42] a saving of just 31 per cent. In 2001 Toyota unveiled its 'Earth-friendly ES3 Concept Car' at the International Frankfurt Motor Show.[43] Though it did not translate this figure into grams of carbon dioxide, it claimed that the ES3 could travel for 100 kilometres on 2.7 litres of fuel. If true, this would have made it twice as energy-efficient as the Prius.* After generating reams of favourable publicity for Toyota, the ES3 was never seen again. It turns out to have been nothing more than an experiment, abandoned, perhaps, because it was insufficiently sexy.

I have an advertisement in front of me extracted from the *Daily Telegraph* of 20 October 1983,† four years after the revolution in Iran raised oil prices to the equivalent of $80 a barrel. After boasting about reaching a 'speed that's well above the legal limit' (the regulation of advertisements has improved a little since then), Peugeot claimed that its new 205 diesels 'are the world's first production cars to do over 72 miles per gallon at a steady 56mph'.[44] This suggests, if true, that a car sold twenty-three years ago was 40 per cent more efficient than the best the mass market has to offer today: the Prius manages just 51 miles per gallon on highways.[45] But the illegal speeds that Peugeot's 205 diesels could reach just aren't high enough for the freedom-seekers of the twenty-first century. Average fuel efficiency in the European Union has improved slowly: by 8 per cent since 1995.‡[46,47] In the United States it has deteriorated. In 1988 the average

---

* Prius: 55 miles per gallon (average across all conditions) = 12 miles per litre = 19.2 km per litre.
ES3: 37km per litre.
† I am indebted to Byron Wine for sending this to me.
‡ From 185 grams of carbon per kilometre to 171.

mileage per gallon* of cars and light trucks was 22.1. Today it is 20.8.[48] This is 17 per cent worse than the Model T Ford, which – in 1908 – managed 25 miles to the gallon.[49] The Ford Expedition, one of the best-selling lines today, achieves 15.5.[50]

So little importance is attached to fuel economy that we struggle even to find reliable figures. In 2005 *Auto Express* magazine conducted a study which found that the British government's mileage calculations for new cars were fictional. 'The official test is carried out on a mechanical rolling road and bears no comparison to real-life driving on UK roads,' the editor of *Auto Express* said. 'Our test team discovered that on average cars are around 17–20 per cent less economical than the official claims.'[51] The same relationship holds in the United States. According to *Detroit News*, 'most drivers achieve only about 75 per cent of the laboratory-generated figures.'[52] In 2006, *Which?* magazine tested the new hybrid cars, which are supposed to be the greenest vehicles on the road. It found that the Toyota Prius managed only 68–76 per cent of its official mileage, while the Honda Civic was only 52–63 per cent as efficient as the UK's Vehicle Certification Agency had claimed.[53]

In 1991, the Rocky Mountain Institute in the United States published a design for a 'Hypercar' which, it claimed, could save at least 70–80 per cent of the fuel other models used.[54] The Institute's critical innovations involved massive reductions in the vehicle's weight and drag. It proposed that the steel body should be replaced with carbon fibre composites, Kevlar or fibreglass and that the underside of the car be made as smooth as its roof. Then, like the Toyota Prius, it would use a 'hybrid-electric drive' (powered by a combination of liquid fuel and an electric motor) which could turn the energy now lost when the car brakes into electricity.

Fifteen years later, though the Institute's design seems to be viable, safe and cheap, and some of its features have been incorporated into real models like the Prius, nothing resembling the whole package has been launched as a mass-market car on either side of the Atlantic. The manufacturers will produce the odd demonstration model – largely, it seems, to keep the regulators off their backs – but while they make

---

* This rather than grams of carbon seems to be the standard comparison in the US.

most of their money on sports utility vehicles, they are simply not interested in serious fuel economies. It is beginning to look like the last days of the Roman empire.

Unwilling to contemplate either a major reduction in the use of the private car or a decline in performance, both the manufacturers and the governments seeking to keep the metaphor alive are now turning instead to alternative fuels. We can keep driving as we do today, they insist, and save the biosphere, by swapping liquid fossil fuels for either biofuels or hydrogen.

Biofuels, in this context, are transport fuels made from plant or animal matter. At first sight this is a beautiful idea. Instead of dragging a filthy black sludge from the bowels of the earth, we would grow our fuel in swaying fields of grain or flowers. It could be made from rapeseed, sunflowers, maize, wheat, sugar cane, even straw or – one day – wood. There would be no more tankers running aground on rocky shores, no more corrupt regimes propped up by black gold, no more murders of indigenous people, no more wars in the Middle East. Most pertinently, cars would produce only the carbon that the plants had absorbed from the atmosphere.

Unsurprisingly, biofuels have caused great excitement among both environmentalists and governments. In May 2005, George W. Bush gave a speech at the Virginia BioDiesel Refinery in West Point.

by developing biodiesel, you're making this country less dependent on foreign sources of oil ... Our dependence on foreign oil is like a foreign tax on the American Dream, and that tax is growing every year ... My administration ... would require fuel producers to include a certain percentage of ethanol and biodiesel in their fuel. And to expand the potential of ethanol and biodiesel even more, I proposed $84 million in my 2006 budget for ongoing research.[55]

This rose to $150 million in the 2007 budget.[56] The 2005 Energy Policy Act obliges fuel companies to sell 7.5 billion gallons of biodiesel and ethanol a year.[57]

On the other side of the Atlantic, our governments have started implementing a similar law. The European biofuels directive rules that 5.75 per cent of our transport fuel should come from renewable sources by 2010.[58] The European Commission intends to raise this to 20 per cent by 2020.[59] The British government has reduced the tax on biofuels by 20 pence a litre, while the European Union is paying farmers an extra 45 euros a hectare to grow them. At last, it seems, a bold environmental vision is being pursued in the world's rich nations.

There are just two problems. The first is the one I mentioned in Chapter 6: we have a finite amount of agricultural land and of the water required to irrigate it. While this limits the production of wood for burning in power stations or boilers, it imposes a far more serious constraint on the cultivation of the starch, sugar or oil required to make liquid fuels. The crops which produce them have to be grown on arable land. When biofuels are widely deployed, they will help precipitate a global humanitarian disaster.

Used on a small scale – as they are now – they are harmless. The people slithering around all day in vats of filth in order to turn used chip fat into motor fuel are performing a service to society. But there is enough waste cooking oil in the United Kingdom to meet only one 380th of our demand for road transport fuel.* The rest has to be grown intentionally.

Road transport in the United Kingdom consumes 37.8 million tonnes of petroleum products a year.[61] The most productive oil crop which can be grown in this country is rape. The average yield is between 3 and 3.5 tonnes per hectare.[62] One tonne of rapeseed produces 415 kilos of biodiesel.[63] So every hectare of arable land could provide 1.45 tonnes of transport fuel. This means that running our cars, buses and lorries on biodiesel would require 25.9 million hectares. You begin to understand the problem when you discover that there are only 5.7 million hectares of arable land in the United Kingdom.[64] Switching to green fuels requires four and half times that. The European Union's target of 20 per cent by 2020 would consume almost all our cropland.

If the same thing is to happen throughout the rich world, the impact could be great enough to push hundreds of millions of people into

---

* The British Association for Biofuels and Oils estimates the volume at 100,000 tonnes a year.[60]

starvation, as the price of food rises beyond their means. If, as some environmentalists demand, it is to happen worldwide, then much of the arable surface of the planet will be deployed to produce food for cars, not people. The market responds to money, not need. People who own cars – by definition – have more money than people at risk of starvation: their demand is 'effective', while the groans of the starving are not. In a contest between cars and people, the cars would win. Something rather like this is happening already. Though 800 million people are permanently malnourished, the global increase in crop production is being used mostly to feed animals: the number of livestock on earth has quintupled since 1950.[65] The reason is that those who buy meat and dairy products have more purchasing power than those who buy only subsistence crops.

The environmentalists who support the wider use of biofuels picture the crops they like best. They see the nodding heads of sunflowers, or the blue blossoms of the linseed plant. They talk of algae which can be grown in desert ponds, or the use of straw and other wastes to produce ethanol. They see towns and villages becoming self-sufficient in fuel, as they are supplied by the farmers down the road. But what they will not see – in fact what they flatly and repeatedly refuse to understand – is that a global commodity market selects not the most satisfying vision, but the cheapest commodity. And at present and for the foreseeable future the cheapest commodity is palm oil. What this means is that biofuel production is a formula not only for humanitarian disaster but also for environmental catastrophe.

In 2005, Friends of the Earth published a report about the impacts of palm oil production.

Between 1985 and 2000, the development of oil-palm plantations was responsible for an estimated 87 per cent of deforestation in Malaysia.[66]

In Sumatra and Borneo, some 4 million hectares of forest has been converted to palm farms. Now a further 6 million hectares is scheduled for clearance in Malaysia, and 16.5 million in Indonesia. Almost all the remaining forest is at risk. Even the famous Tanjung Puting National Park in Kalimantan is being opened up by oil planters. The orang-utan is likely to become extinct in the wild. Sumatran rhinos, tigers, gibbons, tapirs, proboscis monkeys and many other species could go the same way. Thousands of indigenous people have been evicted from their lands, and some 500 Indonesians have been tortured when they tried to resist.[67] The entire region is being turned into a vegetable oil field.

Before oil palms are planted, vast forest trees, containing a much greater store of carbon than the palm trees will ever accumulate, must be felled and burnt. The forest fires which every so often smother the region in smog are mostly started by palm growers. Having used up the drier lands, the plantations are now moving into the swamp forests, which grow on peat. When they've cut the trees, the planters drain the ground. As the peat dries it oxidizes, releasing even more carbon dioxide than the burning trees produce. A paper published in *Nature* estimates that the fires ignited in Indonesia in 1997, which spread as a result of the felling of rainforest trees, released between 13 and 40 per cent as much carbon dioxide as the entire world's consumption of fossil fuels.[68] The biodiesel industry has accidentally invented the world's most carbon-intensive fuel.

The expansion of palm oil plantations is already being driven by the rich world's demand for biodiesel. According to the *Malaysia Star*,

The focus on a new source of demand for crude palm oil – its conversion to biodiesel – has captured the market's imagination. The demand for biodiesel will come from the European Community, where it is mandatory for diesel to be blended with at least 5 per cent of vegetable oil. This fresh demand for crude palm oil would, at the very least, take up most of Malaysia's crude palm oil inventories.[69]

The environmentalist David Bassendine calculates that if 5 per cent of the transport fuel we use in the European Union were carbon-free, this would reduce the world's carbon emissions by a maximum of 0.2 per cent.[70] In other words, the potential carbon saving is a tiny fraction of the possible emissions caused by biodiesel production.

Even when confronted with the construction of nine new palm oil refineries built to meet the European demand for biodiesel,[71] the environmentalists who have embraced this technology still refuse to understand the problem. After drawing attention to the impacts of palm oil production in the *Guardian*, I was bombarded with angry messages insisting that biodiesel could and should be produced at home. But the British government has already investigated and dismissed the possibility of restricting imports. Its report on the subject admitted that

the main environmental risks are likely to be those concerning any large expansion in biofuel feedstock production, and particularly in Brazil (for sugar cane) and South East Asia (for palm oil plantations).[72]

It suggested that the best means of dealing with the problem was to prevent environmentally destructive fuels from being imported. So it asked its consultants whether an import ban would infringe world trade rules. The answer was 'yes': 'mandatory environmental criteria . . . would greatly increase the risk of international legal challenge to the policy as a whole.'[73]

'So change the world trade rules!' the enthusiasts respond. There might be some virtues in this proposal, but if we have to fight for a new global trading regime *before* we can engineer a low-carbon transport system, we might as well give up now. Anyone who has watched the world trade talks – both their snail's progress and the direction in which they are travelling – can see that it would take too long.

The decision by governments in Europe and North America to pursue the development of biofuels is, in environmental terms, the most damaging they have ever taken. Knowing that the creation of this market will lead to a massive surge in imports of both palm oil from Malaysia and Indonesia and ethanol from rainforest land in Brazil; knowing that there is nothing meaningful they can do to prevent them; and knowing that these imports will accelerate rather than ameliorate climate change; our governments have decided to go ahead anyway.

Hydrogen is rather more interesting. Cars, just like our homes, the proponents say, could be powered by hydrogen fuel cells. They would be quiet, they would produce no local pollutants, and they would be around one-third more efficient than internal combustion engines.*

---

* The Tyndall Centre says, 'Fuel cells can deliver a sustained 60 per cent energy conversion efficiency, whereas ICEs [internal combustion engines] have a maximum efficiency of 40 and 45 per cent for petrol and diesel respectively, and normally operate well below this level.'[74]

The hydrogen, again, could be made from gas or coal (and the carbon dioxide produced in its manufacture buried) or from the electrolysis of water by means of renewable energy.

The European Commission has already devoted €2 billion to this project,[75] and the Bush administration has more or less matched it.[76] The US National Academy of Engineering estimates that 60 per cent of the new vehicles sold in the United States in the year 2034 could be run on hydrogen fuel cells, though it admits this is an 'optimistic vision'.[77] The Japanese government says it wants 50,000 hydrogen vehicles on its roads by 2010, and five million by 2020.[78] A report by the Tyndall Centre, released in 2002, notes that

The major motor manufacturers are predicting that hydrogen fuel cell vehicles will be made available on a limited basis by 2004, and hydrogen-powered internal combustion engines in the next 5–7 years.[79]

The manufacturers' claims were nonsense. But hype like this is often swallowed whole by people who ought to know better. In their book *Natural Capitalism*, published in 1999, Paul Hawken and Amory and Hunter Lovins – seeking to demonstrate that their ideas were being adopted by the automakers – reported that

In April 1997, Daimler-Benz announced a $350 million joint effort with the Canadian firm Ballard to create hydrogen-fuel-cell engines. Daimler pledged annual production of 100,000 such vehicles per year by 2005, one seventh of its total current production. Six months later, the president of Toyota said he'd beat that goal, and predicted hybrid-electric cars would capture one third of the world car market by 2005 . . . by the spring of 1998, at least five automakers were planning imminent volume production of cars in the 80 mpg range.[80]

These authors might like to reflect on the fact that in 1929 the president of General Motors predicted that his cars would be managing 80 miles to the gallon within ten years.[81] The motor industry makes grand predictions of this kind every few months, in order to excite its investors and show that it intends to save the planet. We would be well advised not to believe a word it says.

Even so, hydrogen fuel cells are beginning to look like a feasible technology for motor transport, if not on the timescale the producers predict. Before they can be widely deployed, they will have to negotiate several major obstacles.

The most immediate problem is that hydrogen cannot be bought in filling stations. The chicken and egg problem I described in Chapter 7 is even harder to negotiate for mobile fuel cells. Stationary models will be supplied from just one pipe. The owners of fuel cell cars need to be sure that they can find hydrogen wherever they happen to run out. The filling stations won't supply it until they have a market, and the market can't develop until there are supplies.

This is compounded by the problem of storage, which is not something the owners of stationary fuel cells need to worry about unless they produce their own hydrogen. Cars would need to take it with them. Though the gas is three times as energy-dense as petrol in terms of weight, it is only one tenth as dense in terms of volume – at pressures of 5,000 pounds per square inch.*

This means that a hydrogen-powered vehicle would need a high-pressure fuel tank ten times the size of a petrol-driven car's in order to travel as far. High-pressure tanks would take a long time to fill, and could be dangerous. Alternatively, the hydrogen could be liquefied, but this, as I mentioned in Chapter 7, requires a great deal of energy. As soon as the car is parked and the engine has stopped running, it will start to boil off. According to the US National Academy of Engineering,

New solutions are needed in order to lead to vehicles that have at least a 300-mile driving range ... no hydrogen storage system has yet been developed that is simultaneously lightweight, compact, inexpensive, and safe.[83]

One of the new solutions the automakers have attempted is 'on-board re-formation' of hydrogen. What this means is that the cars contain a miniature factory. Their fuel tanks are filled with either petrol or methanol and the re-former turns it to hydrogen while they are on the road. But on-board re-formers are expensive, heavy, inefficient and take a long time to start.[84] Most importantly, they can make no

---

* Hydrogen contains 3 megajoules of energy per litre at 5000 psi. Petrol contains 32 megajoules at atmospheric pressure.[82]

contribution to tackling climate change, as the production of hydrogen can neither be powered by renewable energy nor accompanied by carbon burial. The only benefits are that the cars are quieter and produce less local pollution. All these problems, of course, also attend the development of hydrogen-powered internal combustion engines.

The only remaining option seems to be storage in the form of a solid of some kind. Hydrogen can be adsorbed by carbon or used to create metal hydrides, which release the gas (and a good deal of heat) when reacted with water. This approach is also beset by problems, such as the likely leakage of hydrogen and the temperatures at which the reactions take place.[85] The manufacturers have so far been unable to show how the spent metal hydrides in the car can be extracted and recycled. But experimental vans running on sodium borohydride have already been tested by DaimlerChrysler. They appear to be light enough and to have a long enough range for commercialisation, assuming that the other problems can one day be overcome.

Then there's the cost. The Tyndall Centre reports that 'Fuel cells for vehicles currently cost around $1,000 per kilowatt whilst internal combustion engines cost $10 per kilowatt.'[86] The National Academy says the whole system

will have to decrease in cost to less than $100 per kilowatt before fuel cell vehicles become a plausible commercial option ... it will take at least a decade for this to happen.[87]

But while the system is expensive, the fuel is not. When oil cost $30 a barrel, the Academy's charts show, a kilogram of hydrogen produced from gas or coal, even when the carbon dioxide is buried, was only a little more expensive than a kilogram of gasoline.[88] At the time of writing, oil costs about twice that amount, which means that hydrogen is currently cheaper than petrol. This is not the case, however, when it is produced by electrolysis.

While hydrogen cars could one day make the world's oilrigs redundant, they would require a massive increase in other kinds of energy production. In Europe, the fuel is likely to be produced from natural gas, which would, of course, hasten its eventual decline. In North America, whose gas reserves are in freefall, it will be made from coal or (more expensively) electrolysis using nuclear power or wind. This would require roughly a doubling of national electricity production.[89,90]

But the fundamental problem with cars powered by hydrogen fuel cells is that – because of the unresolved technical problems I've mentioned – their development will simply be too slow, despite the billions being thrown at them by the rich world's governments. The National Academy's 'optimistic vision' looks like nothing more than wishful thinking by the time you have read the whole report. It admits that

> although a transition to hydrogen could greatly transform the US energy system in the long run, the impacts on oil imports and carbon dioxide emissions are likely to be minor during the next 25 years.[91]

This assessment seems to be shared by *New Scientist* magazine, which in 2003 reported that 'It will probably be 15 to 20 years before fuel-cell cars gain even a toehold in the market.'[92]

While mobile fuel cells might one day change the world, in other words, as far as my target is concerned they are pretty well useless. In lobbying for lower emissions, our time would be better spent demanding the mass production of a hypercar – which faces no significant technical barriers – rather than the use of hydrogen.

Alternatively, there might be a means of overcoming the main drawback of the electric vehicle. Electric cars already exist, and some new models are as fast as any vehicle needs to be. But its range is limited by the capacity of the battery: the best ones run out after 100–300 miles and take hours to recharge. The energy expert Dave Andrews suggests a simple solution: the cars should use batteries provided by a network of filling stations. As the battery runs down, you pull into a station, pay a fee and swap it for another one. The stations could charge their batteries from electricity provided by off-shore wind farms when the wind is blowing strongly and demand is low. This means that the carbon costs would be roughly zero, surplus wind power would not be wasted and the financial costs would remain small, as the power is bought when it is cheap. This looks like a far simpler and less costly means of reducing the carbon content of transport fuel than a hydrogen network. Given that the alternatives are so much easier to develop, our governments' obsession with hydrogen cars seems incomprehensible.

In her excellent book *Car Sick*, the transport analyst Lynn Sloman, drawing on surveys conducted in Australia, three English towns and a rural area in mid-Wales, derives what she calls 'the 40:40:20 rule'.[93] Irrespective of location, some 40 per cent of current car journeys could already be made by bicycle, on foot or by public transport. Another 40 per cent could be made by other means if public transport or cycling facilities were improved. Roughly 20 per cent of car journeys cannot be swapped.

Most of the means of persuading drivers to use other modes of transport are – by comparison to the billions spent on building roads and bridges – simple and cheap. One of the biggest problems, Sloman found, is that people don't know about existing services. By visiting people's homes and handing out introductory free tickets, bus companies in some parts of Britain have been able greatly to increase their custom. Advertising can help to correct people's misconceptions about the alternatives: one survey showed that people overestimate the time a journey would take by public transport by 70 per cent and underestimate the time it would take by car by 26 per cent.[94]

Local bus services can also be greatly improved. In Germany, Holland and Denmark, the regional or provincial governments determine which services are needed, and require the bus companies to sign a contract to provide them. In the United Kingdom, by contrast, it is left to the market. The companies ply the most lucrative routes at the most lucrative times, and park the buses before they have to pay overtime to the drivers. The idea of 'cross-subsidization' (the profits from the busy routes help to pay for services on the quieter ones) is much discussed in this country but seldom imposed.

Our public transport systems are blighted by a lack of imagination. While a bus service in the United Kingdom means just one thing – a bus trundling up and down a fixed route at set times – in parts of Holland, Germany, Switzerland and Denmark, it describes several different kinds of transport. There are, for example, taxi-buses, using vehicle tracking systems, global positioning satellites and call centres. These are a kind of shared taxi service, which you book in advance and which comes to your door, but which picks up other passengers – who have also made bookings – along the way. There are 'bell buses', which run on quiet routes to regular timetables, but only if someone has already phoned to request the service.[95] This enables them to cut both costs and emissions.

Other European countries – especially Holland, Germany and Switzerland – also possess integrated transport services of the kind the United Kingdom was promised in 1998,[96] but which still hasn't materialized. Buses are scheduled to meet trains. Trains, trams and buses carry bicycles, free of charge and at most times of the day. Safe, continuous cycle lanes connect with each other and traverse entire cities.

In a few towns in the United Kingdom, parents take turns to escort 'walking buses' – crocodiles of children – to school. By obliging children to use appendages which have otherwise become almost vestigial – their legs – they reduce both traffic and obesity. Because only two adults – a 'driver' and a 'conductor' – are needed to herd the children, their parents don't have to travel to school every day.

None of this is either expensive or difficult to implement, but most local and national governments in the United Kingdom, like those of the United States, are just not interested. They are obsessed with congestion – with the impediments, in other words, to motorists' perceived freedom – and seek to reduce it by creating more space for the car.

Another means of cutting carbon emissions without the need for new technologies is to reduce our need to travel. In Chapter 9 I propose a shopping system which more or less eliminates the requirement for private cars, with an associated carbon cut of around 70 per cent. There is also scope – though not as much as some of its boosters have suggested – for reducing the need to travel to work.

Studies in California, the Netherlands and Germany suggest that people who are able to work at home for part of the week cut the total distance they travel, for all purposes, by between 10 and 20 per cent.[97] This might be offset a little by the fact that they could be tempted – if they don't have to go in so often – to live further away.

By 2001, 2.2 million people in the United Kingdom, or 7.4 per cent of the workforce, worked from home at least one day a week.[98] The government's advisers suggest that if the number continues to grow at the current rate, in ten years around 30 per cent of the country's workforce will be teleworking for at least some of the time.[99]

But the growth of teleworking can't go on for ever. It is hard to see how factory workers, shop assistants, nurses, drivers, builders or gardeners could take much of their work home. The advisers point out that the jobs which can most easily be taken home (those which involve computers and telephones) are the ones which have been growing fastest. They forget to mention, however, that these are also the sectors which are most likely to disappear overseas: if your boss can allow you to work from home, he can also send your job to Hyderabad.

A study in the United States suggests that 50 per cent of employees there are 'information workers', and 80 per cent might be able to work from home, which would mean that the maximum proportion of teleworkers would be about 40 per cent.[100] If everyone who was able to do so spent two days a week teleworking, this would cut the number of commuting journeys by 16 per cent. This is optimistic. An opinion poll in the UK suggests that of those who use computers but don't yet work from home, 77 per cent don't want to do it, 17 per cent would like to do it but wouldn't be allowed to, and only 7 per cent would and could.[101] But a rationing system would change these proportions, and video broadband might help to counteract the feeling of isolation that often discourages peoples from staying at home.

There is also some potential for sharing journeys. Several companies and councils in the UK already encourage their workers to share their cars, or round them up in minibuses. One study suggests that an average of 14 per cent of the workers in the companies running car-sharing schemes are making use of them.[102] Two firms – Marks and Spencer Financial Services and Computer Associates, both of which give money to people who share – have persuaded around one third of their staff to join the system.[103]

There are also car-sharing schemes for people who don't work together. To use the Liftshare system, for example, you simply post the journey you intend to make and the date on which you wish to travel on its website and say whether you are offering or seeking a lift. The scheme will attempt to find someone else who is going the same way.[104] The site will tell you how much the journey costs, and how many kilograms of carbon dioxide you save by travelling with someone else. In other words, it's a form of hitch-hiking that relieves you of the need to stand on the hard shoulder in the rain being abused by Volvo drivers. It's also likely to be safer. Liftshare claims to have 104,000 members, who share just over 41 million kilometres of journeys a year. It says 32 per cent of its requests lead to a successful match.[105] All these numbers – for both workplace and public schemes – would be greatly boosted by a rationing system, which would encourage people to share the carbon costs of travel.

Many of these proposals are likely to be as effective at glueing communities back together as they are at reducing carbon emissions. Just before I wrote this chapter, I counted the number of people I passed on the pavement when walking from my house to the local bakery, four streets away. Out of sixteen, I knew twelve by name. It was a slow journey, but I cannot think of a better way of spending my time.

It is not hard to see how a universal switch to hypercar technologies or electric vehicles and a return to lower speeds and lower standards of performance, accompanied by car sharing, tele-commuting, a car-free shopping scheme, better public transport and better facilities for cyclists and walkers, could cut emissions by more than 90 per cent across the journeys that Storkey's system could not replace. But the problem is political, not practical. We need governments to start deciding how best to run a transport system, rather than how best to accommodate the private car. That means confronting a lobby which appears to become more confident by the year, as the libertarian politics encouraged by driving make any limitations on drivers harder to achieve. The longer governments prevaricate, the less plausible substantial change becomes; the problem which in other respects is the easiest in this book to fix is in danger of becoming insoluble. One of our greatest political challenges is to prove Ilya Ehrenburg wrong.

9

Love Miles

Dismiss my aerial engine, which on cloudless days
Has spirited me gently over land and sea.

Faust, Part II, Act IV[1]

Every so often, I receive an e-mail from a company called respon-
sibletravel.com, advertising 'Holidays that give the world a break'.[2]
It arranges 'real and authentic holidays that also benefit the environment and local people'.[3] You can travel to the Quelqanqa Valley in
Peru, where you can help build 'three small footbridges across the
two rivers that run close to the village'.[4] You can take 'a one-week
safari with a difference!' in Kenya, where the fees you pay to enter
the Samburu National Park and the Maasai Mara reserve fund
'amenities such as schools and medical facilities for the local people'.[5]
You can visit Hokkaido in Japan, where your holiday will help to
restore 'original wetland'.[6] Its website quotes the Cree Indian saying
reproduced endlessly on t-shirts and posters: 'Only when the last tree
has died and the last river been poisoned and the last fish been caught
will we realize that we cannot eat money.'[7]

Even if we forget that Andean villagers are likely to be rather better
at building bridges than IT consultants from north London, that the
gate fees from Kenyan game reserves tend to be spent on anything
except amenities for local people,[8] that it would be hard to charac-
terize the Japanese as being in need of foreign aid and that – to judge
by its tariffs – responsibletravel.com is just as good at eating money
as any other travel firm, something is missing from its account. The
tourists have to get there. And unless the holiday incorporates a bike

ride to the Maasai Mara or a rowing trip to Hokkaido, they are going to get there by air.

You could build 3,000 footbridges, spend your life's savings on gate fees in Kenya, slosh around in wetlands until you had trench foot, and not redress a fraction of the impact caused by your flight. In the name of assisting the people of developing countries, this company is helping to starve the Ethiopians and drown the people of Bangladesh.

I have picked on responsibletravel.com not because it is exceptional in its disregard for the impact of air travel, but because it is typical. We have all mastered the art of beginning and ending a narrative at the points which suit us, and we are never more adept at this than when we travel. A holiday – and therefore its environmental impact – we choose to believe, begins upon arrival and ends upon departure. The getting there and back has nothing to do with it.

Even those who recognize that we don't arrive by means of teleportation seek to underplay the impacts of aviation. In an article in the *Independent* extolling the virtues of 'ethical tourism', Anita Roddick, founder of the Body Shop and funder of a thousand good causes, acknowledges that

Some people are suspicious of the whole idea of ethical tourism, arguing that, to reach most places, you have to travel by air – which is itself unethical. It is true that the government expects air travel passengers to double by 2030, by which time air travel will be the biggest contributor to global warming. However, it is also true that the airline industry can now create fuel-efficient planes. The Sustainable Aviation Group, which includes BA and Virgin, aims to introduce new aircraft producing 50 per cent less carbon dioxide than 2000 models.[9]

This, as we shall see, is unlikely to happen: the airline companies' projections resemble those of the motor industry, in that they appear to be designed principally for the purpose of public relations. Even if this improvement did take place, it would not counteract the rising emissions caused by the growth in flights, as Roddick herself shows. She is also sufficiently well informed to know that the other measures she proposes – emissions trading and tree planting – are not, under current circumstances, going to work, as I will explain in a moment and in Chapter 11. But her determination to suggest that long-distance tourism is sustainable as long as we distribute money and goodwill when we arrive supports my contention that well-meaning people are as capable of destroying the biosphere as the executives of Exxon.

Our moral dissonance about flying reminds me of something a Buddhist once told me when I questioned his purchase of unethical products. 'It doesn't matter what you do, as long as you do it with love.' I am sure he knew as well as I did that our state of mind makes no difference either to the exploitation of workers or to the composition of the atmosphere. Thinking like ethical people, dressing like ethical people, decorating our homes like ethical people makes not a damn of difference unless we also behave like ethical people. When it comes to flying, there seems to be no connection between intention and action.

This is partly, I think, because the people who are most concerned about the inhabitants of other countries are often those who have travelled widely. Much of the global justice movement consists of people – like me – whose politics were forged by their experiences abroad. While it is easy for us to pour scorn on the drivers of sports utility vehicles, whose politics generally differ from ours, it is rather harder to contemplate a world in which our own freedoms are curtailed, especially the freedoms which shaped us.

I have heard people argue that less travelling from the rich nations to the poor nations could result in a narrowing of the public mind. This might be true. But it is also clear – as the public response to the Asian tsunami and the recent enthusiasm for tackling poverty in Africa suggests – that our compassion for other people can be stimulated just as well by effective use of the media.

More painfully, in some cases our freedoms have become obligations. When you form relationships with people from other nations, you accumulate love miles: the distance between your home and that of the people you love or the people they love. If your sister-in-law is getting married in Buenos Aires, it is both immoral to travel there – because of climate change – and immoral not to, because of the offence it causes. In that decision we find two valid moral codes in irreconcilable antagonism. Who could be surprised to discover that 'ethical' people are in denial about the impacts of flying?

There are two reasons why flying dwarfs any other impact a single person can exert. The first is the distance it permits us to cover. According to the Royal Commission on Environmental Pollution, the carbon emissions per passenger mile 'for a fully loaded cruising airliner are comparable to a passenger car carrying three or four people'.[10]

In other words, they are about half those, per person, of a car containing the average loading of 1.56 people. But while the mean distance travelled by car in the United Kingdom is 9,200 miles per year,[11] in a plane we can beat that in one day. On a return flight from London to New York, every passenger produces roughly 1.2 tonnes of carbon dioxide:* the very quantity we will each be entitled to emit *in a year* once a 90 per cent cut in emissions has been made.†

The second reason is that the climate impact of aeroplanes is not confined to the carbon they produce. They release several different kinds of gases and particles. Some of them cool the planet, others warm it. The overall impact, according to the Intergovernmental Panel on Climate Change, is a warming effect 2.7 times that of the carbon dioxide alone.[13] This is mostly the result of the mixing of hot wet air from the jet engine exhaust with the cold air in the upper troposphere, where most large planes fly. As the moisture condenses it can form condensation trails which in turn appear to give rise to cirrus clouds – those high wispy formations of ice crystals sometimes known as 'horsetails'. While they reflect some of the sun's heat back into the space, they also trap heat in the atmosphere, especially at night. The heat trapping seems to be the stronger effect.[14] This means that subsonic aircraft, if all their seats are full, cause roughly the same total warming per passenger mile as cars. While the different warming effects are not directly comparable, because carbon dioxide stays in the atmosphere for much longer than condensation trails or cirrus clouds, if we were to multiply the carbon emissions produced on that round-trip to New York by 2.7, we would, of course, exceed our annual allowance on that journey by the same factor.

---

* The UK Department for Transport says that long-haul flights produce 110 grams of carbon dioxide per passenger kilometre.[12] New York is 5585 kilometres from London.

† 0.33 tonnes of carbon × 3.667.

Supersonic aircraft, such as Concorde (which is now retired) and some military planes, are far more damaging. They fly not in the upper troposphere (where planes cruise at between 10 and 13 kilometres), but in the stratosphere, at between 17 and 20 kilometres above the surface of the earth. The water vapour they produce there ensures that their total impact is around 5.4 times that of the carbon dioxide alone. Discussing the small supersonic 'business jets' whose development is allegedly being pursued by NASA, General Electric and Lockheed Martin,[15] the Royal Commission abandons its customary restraint.

The contribution to global climate change of this kind of aircraft would be so disproportionate that their development and promotion must be regarded as grossly irresponsible.[16]

Aviation has been growing faster than any other source of greenhouse gases. Between 1990 and 2004, the number of people using airports in the United Kingdom rose by 120 per cent, and the energy the planes consumed increased by 79 per cent.[17] Their carbon dioxide emissions almost doubled in that period – from 20.1 to 39.5 million tonnes,[18] or 5.5 per cent of all the emissions this country produces.[19]

Unless something is done to stop this growth, aviation will overwhelm all the cuts we manage to make elsewhere. The government predicts that, 'if sufficient capacity were provided', the number of passengers passing through airports in the United Kingdom will rise from roughly 200 million today to 'between 400 million and 600 million' in 2030.[20] It intends to ensure that this prophecy comes to pass. The new runways it is planning 'would permit around 470 million passengers by 2030'.[21]

You might wonder how the British government reconciles this projection with its commitment to cut carbon emissions by 60 per cent by 2050. The answer is that it doesn't have to. As the Department for Transport cheerfully admits,

International flights from the UK do not currently count in the national inventories of greenhouse gas emissions as there is no international agreement yet on ways of allocating such emissions.[22]

This is a remarkable evasion. It is true that there is 'no international agreement yet'. But a child could see that you simply divide the emissions by half. The country from which passengers depart or in which they arrive accepts 50 per cent of the responsibility. Are we really to believe that the civil servants in the Department for Transport can't work this out? As 97 per cent of the greenhouse gases they expect planes to be producing by 2030 will come from international flights,[23] this profession of incompetence is, to say the least, convenient. You need do nothing about the carbon emissions from aeroplanes as officially they don't exist.

By way of remedy, the transport department suggests that the aviation industry should 'pay the external costs its activities impose on society at large'.[24]

This is an interesting proposal, but unfortunately the department does not explain how it could be arranged. Should a steward be sacrificed every time someone in Ethiopia dies of hunger? As Bangladesh goes under water, will the government demand the drowning of a commensurate number of airline executives? The idea is strangely attractive. But the only suggestion it makes is that aviation fuel might be taxed:

a notional 100 per cent fuel tax would lead to . . . a 10 per cent increase in air fares, assuming the increased costs were passed through in full to passengers. This would then have the effect of reducing demand by 10 per cent.[25]

A few pages later, it admits that this mechanism is in fact useless, because the airlines will keep cutting the remainder of their costs.[26] The government is also aware that aviation fuel taxes on international flights are more or less impossible to impose. They are prohibited under international law by Article 24 of the 1944 Chicago Convention, which has been set in stone by around 4,000 bilateral treaties.[27] We environmentalists have been stupid enough to do what the industry wants, and loudly demand the taxes it has no cause to fear.[28]

So the government relies instead on incorporating aviation into the European Emissions Trading Scheme. This, it hopes, will happen in 2008. In principle – though with the major caveats I mentioned in Chapter 3 – the idea is sound: an overall carbon limit is set for the participating industries, and the market is left to allocate emissions between them. The problem is that if government policy is still driving the growth of the airlines, and low prices continue to stimulate demand, either the emissions from every other industry within the scheme must contract at a much greater rate than before to accommodate aviation's expansion, or it will break the system. The second option appears to be the more likely. Incorporating the industry into the trading scheme without other policies to reduce its growth merely defers the decision the government needs to take, while threatening the remainder of its climate-change programme.

The one certain means of preventing the growth in flights is the one thing the British government refuses to do: limit the capacity of our airports. It employs the 'predict and provide' approach which has proved so disastrous when applied to road transport:* as you increase the provision of space in order to meet the projected demand, the demand rises to fill it, ensuring that you need to create more space in order to accommodate your new projections. The demand would not have risen in the first place if you hadn't created the space. The House of Commons Environmental Audit Committee calculates that the extra capacity the government proposes means 'the equivalent of another Heathrow every 5 years'.[30] Twelve regional airports in the UK have recently announced expansion plans.[31] Ministers are now beginning to promote new runways at Heathrow, Stansted, Birmingham, Edinburgh and Glasgow.

In 2005, Friends of the Earth asked the Tyndall Centre to determine what impact this growth would have on greenhouse gas emissions.[32] The results were staggering. If we attempt to stabilize carbon dioxide concentrations in the atmosphere at 550 parts per million (which roughly corresponds to the government's target), and aviation continues to grow as the government envisages, by 2050 it would account for 50 per cent of our carbon emissions. If we tried to stabilize them at 450 parts (which is closer to my target) flying would produce 101 per cent of the carbon the entire economy was able to release. If the

---

* The Environmental Audit Committee notes that, 'Despite protestations to the contrary, it is abundantly clear that the aviation White Paper adopts a "predict and provide" approach.'[29]

carbon emissions were multiplied by 2.7, to take into account the full impact of aviation on the climate, the figures would be 134 per cent and 272 per cent respectively.*[34] The researchers assumed that the fuel efficiency of aircraft will improve by 1.2 per cent a year throughout this period. This could be optimistic.[35]

While the British government appears determined to turn this country into the nation Orwell envisaged in *1984* – Airstrip One – aviation is booming everywhere. Worldwide, it has been growing by about 5 per cent a year since 1997.[36] The Intergovernmental Panel on Climate Change suggested it would account for between 3 and 10 per cent of global carbon emissions by 2050[37] (and that this impact could be amplified 2.7 times). But the Royal Commission reports that growth has so far been higher than it envisaged: the panel's prediction 'is more likely to be an under-estimate rather than over-estimate'.[38]

Faced with both their lobbying power and the aspirations of their customers, hardly any government appears to be brave enough to stand up to the airlines. The British Department for Transport, like the airline industry, claims that expanding airport capacity is 'socially inclusive', in that it enables poorer people to fly.[39] But as the Environmental Audit Committee points out, it seems to have conducted no research on this subject.[40] An organization which has – the Civil Aviation Authority – found that people in social classes D and E (at the bottom of the official economic scale) scarcely fly at all. Though flights are often very cheap, they can't afford to take foreign holidays: even in the age of the 50p ticket, people in these classes buy just 6 per cent of the tickets.[41] A MORI poll commissioned by the Freedom to Fly Coalition (which is a lobby group founded by the aviation industry) found that 75 per cent of those who use budget airlines are in social classes A, B and C.[42] Another survey shows that people with second homes abroad take an average of six return flights a year.[43] But even if everyone in the rich nations were able to fly every year, the impact of aviation would still be regressive, as the people who are

---

* The authors warn that 'Although percentages have been included for an uplift factor of 2.7, it should be noted that there is substantial scientific uncertainty relating to both the size of the factor that should be used, as well as to the method of simply "uplifting" carbon values for comparison with carbon emissions profiles. Strictly speaking, such a comparison does not compare like with like.'[33]

most vulnerable to climate change are the poorest inhabitants of the poorest nations, the great majority of whom will never board an aeroplane.

There are two means by which the growth in flights could be reconciled to the need to cut carbon emissions. The first is a massive increase in the fuel efficiency of aircraft; the other is a new fuel.

The British government's White Paper on aviation claims that

Research targets agreed by the Advisory Council for Aeronautical Research in Europe suggest that a 50 per cent reduction in carbon dioxide production by 2020 can be achieved.[44]

This statement, as the House of Commons Environmental Audit Committee has pointed out, is deliberately misleading.[45] What the Advisory Council actually said is that its target, which is purely aspirational, cannot be met by improving the existing kinds of aircraft engine. It requires 'breakthrough technologies' which don't yet exist.[46] When you consider the design life of modern aircraft, you discover that the council's 'research target' bears even less relation to actual performance. Planes are remarkably long-lived. The 747 – the Jumbo jet – was launched in 1970 and is still flying today. The Tyndall Centre predicts that the new Airbus A380 will still be in the air, though 'in gradually modified form', in 2070,[47] and it will continue to use 'high-pressure, high-bypass jet turbine engines that contain only incremental improvements over their predecessors'.[48]

A 50 per cent cut by 2020 means not only discovering a new technology and designing, testing, licensing and manufacturing the planes that use it, but also scrapping and replacing the entire existing fleet, and the tens of billions of pounds the aviation companies have invested in it.

As far as aircraft engines are concerned, 'breakthrough technologies' appear to be a long way off. The Royal Commission reports that

The basic gas turbine design emerged in 1947. It has been the dominant form of aircraft engine for some 50 years and there is no serious suggestion that this will change in the foreseeable future.[49]

It is hard to see how major new efficiencies could be squeezed out of it. The proposals the aviation industry has put forward, the Royal

Commission says, might improve fuel efficiency somewhat, but only at the cost of an increase in noise and local pollution caused by nitrogen oxides.[50] I don't believe that those who live under the flight paths (an increasing proportion of the population as aviation expands) will put up with this.

Worse still, as a report for the European Commission by the aviation scientist Ulrich Schumann notes,

Recent experiments have provided evidence that contrails form at lower altitude and hence more frequently when using more efficient engines.[51]

'Contrails' are the condensation trails largely responsible for boosting the impact of aviation on the climate, as discussed above. In other words, jet engines might be able to burn less fuel, but the warming they cause could remain constant or even increase.

The only technology which does offer a major improvement in fuel efficiency, creates fewer condensation trails (because it is used at lower altitudes) and is known to work is one plucked not from the future but the past: the propeller plane. According to the industry coalition Avions de Transport Régional, the most efficient commercial propellor planes use just 59 per cent as much fuel per passenger mile as a jet aircraft.*[52] But as the 'Régional' part of their name suggests, they advocate its use only for short-haul flights. Short-haul flights are inherently inefficient, because of the higher proportion of the journey spent taking off and gaining height. They are also, on the whole, unnecessary, as there are other means of covering that distance: it would be better for the environment to travel by coach or by train. The table below gives the figures I provided in Chapter 8, with the carbon emissions from short-haul flights added.

No one in the industry appears to be giving serious consideration to the idea of returning to long-haul propellor planes, because they are much slower than jets. The most promising approach to redesigning the rest of the aircraft is a concept called the 'blended wing-body'. Planes of this kind would have huge hollow wings in which some of the passengers would sit. By cutting drag, they could reduce the amount of fuel a plane uses by up to 30 per cent.[53] But, as the Royal

---

* 16 litres per 200 nautical miles, rather than 27.

mode of transport from London to Manchester	carbon dioxide emissions per passenger (kg)
plane[i] (70 per cent full)	63.9
car (1.56 passengers)	36.6
train (70 per cent full)	5.2
coach (40 passengers)	4.3

i. The Department for Transport estimates that short-haul flights use 150g of carbon dioxide per passenger kilometre[54] Manchester is 298km from London. This gives 44.7kg/ passenger for a fully laden plane.

Commission points out, it's still just a concept, and 'the stability and controllability of such an aircraft are unproven.'[55]

There is some scope for reducing the amount of fuel that planes burn by improving air traffic control (so that they spend less time in the stack) and allowing them to take more direct routes. But this amounts to just 10 per cent or so.[56] If aircraft flew at lower altitudes, they would produce fewer condensation trails and cirrus clouds, but because the air is denser there – so the planes would be subject to more drag – they would burn more fuel.[57] The net effect might be beneficial, but this remains uncertain.[58]

The choice of alternative fuels for aeroplanes is similar to the choice of alternative fuels for cars. According to a paper by researchers at Imperial College, London, it is technically possible to fly planes whose normal fuel (kerosene) is mixed with small amounts of biodiesel.[59] At low temperatures, oils go cloudy, and at a couple of degrees lower still they form a gel. This can block an engine's fuel filters, fuel lines and plugs. Biodiesel's 'cloud point' is much higher than kerosene's. Even a mixture containing as little as 10 per cent biodiesel can raise the cloud point from −51° to −29°. This, because of the low temperatures in the upper troposphere, could stop the engines if the plane flew at normal heights. But if the fuel is repeatedly chilled and the crystals which form filtered out, a 10 per cent mixture raises the cloud point by only 4° (they don't say how much energy this chilling would use).[60] This would permit the plane to fly at up to 9,500 metres. Of course the biodiesel used in planes is subject to the same environmental constraints as the biodiesel used in cars: it is likely to cause more global warming than it prevents.

Ethanol, the same paper suggests, would be useless: it is insufficiently dense and, in aeroplanes, extremely dangerous. Kerosene could be made from wood, but (aside from the enormous expense), its production will be limited by the land-use issues I discussed in Chapter 6: the trees we are able to grow would be better employed discharging the more necessary function of keeping us warm. You will have guessed by now that this leaves only our familiar fall-back, hydrogen.

In this case, it would be burnt not in fuel cells, but in combustion engines similar to those used in planes today. Carrying liquid hydrogen seems to be a more viable option for planes than for cars, but the energy costs (about 35 per cent) of keeping the temperature below −259° remain unchanged. Several planes have already been flown with one engine running on hydrogen. In principle, jets could use this fuel today, if instead of carrying passengers and freight they carried only hydrogen. Though it is lighter, it contains four times less energy by volume than kerosene. But if this problem could be overcome, the researchers at Imperial College suggest, the total climate impacts of planes fuelled by the gas 'would be much lower than from kerosene'.[61] Unfortunately, they appear to have forgotten something.

When hydrogen burns, it creates water. A hydrogen plane will produce 2.6 times as much water vapour as a plane running on kerosene. This, they admit, would be a major problem if hydrogen planes flew as high as ordinary craft. But if, they suggest, the aircraft flew below 10,000 metres, where condensation trails are less likely to form, the impact would be negligible. What they have forgotten is that because hydrogen requires a far bigger fuel tank than kerosene, the structure (or 'airframe') of the plane would need to be much larger. This means it would be subject to more drag. The Royal Commission on Environmental Pollution – which as usual appears to have thought of everything – points out that 'the combination of larger drag and lower weight would require flight at higher altitudes' than planes fuelled by kerosene.[62] In fact, hydrogen planes, if they are ever used, are most likely to be deployed as supersonic jets in the stratosphere. This would be an environmental disaster.

a hydrogen-fuelled supersonic aircraft flying at stratospheric levels would be expected to have a radiative forcing [which means a climate-changing effect] some 13 times larger than for a standard kerosene-fuelled subsonic aircraft.[63]

And that, I am afraid, is it. As the Intergovernmental Panel on Climate Change discovered,

there would not appear to be any practical alternatives to kerosene-based fuels for commercial jet aircraft for the next several decades.[64]

Even the British government, which at other times manages to find its way to the conclusions the aviation industry requires, admits that 'there is no viable alternative currently visible to kerosene as an aviation fuel.'[65]

There is, in other words, no technofix. The growth in aviation and the need to address climate change cannot be reconciled. Given that the likely possible efficiencies are small and tend to counteract each other or to be unacceptable for other reasons, a 90 per cent cut in emissions requires not only that growth stops, but that most of the planes which are flying today are grounded. I recognize that this will not be a popular message. But it is hard to see how a different conclusion could be extracted from the available evidence.

The obvious next question, then, is this. Are there other means of covering the same distances at the speeds with which we are now familiar?

Commercial airliners such as the Boeing 747 or the Airbus A321 have cruising speeds of around 900 kilometres per hour. The fastest form of mass transit across the surface of the earth is the ultra-high-speed train. The French TGV – *Train à Grande Vitesse* – holds the record for a wheeled train, of 515 kmph.[66] Locomotives that are suspended above the track by magnetic repulsion – maglevs or magnetic levitation trains – can beat this. In 2003, a test train in Japan managed 581 kmph.[67] But the fastest working train in the world is the TGV from Lyons to Aix-en-Provence. It covers 290 kilometres in 66 minutes, an average speed of 263 kmph.[68] Trains are not – or not yet – as fast as planes, but when the check-in, boarding and waiting times and the travel to and from the airports are taken into account, they can cover journeys of a few hundred kilometres in roughly the same time. Given that they are generally more convenient and relaxing than planes, it might be possible to persuade people of the advantages of using ultra-high-speed trains for journeys of up to about 2000 kilometres, even though these would take longer. Beyond that point, the journey, for people accustomed to moving at Faust's speed, would start to drag: 8,000 kilometres – from London to Beijing for example – takes 31 hours at 260 kmph. The companies and governments proposing new ultra-fast lines hope to draw people away from planes by raising average speeds to 350 or even 500 kmph.[69,70,71]

The first obstacle you discover is the cost. The 30-kilometre maglev system between Shanghai and its airport cost $1.2 billion.[72] The link between Edinburgh and Glasgow, on which the bullet train some politicians have been proposing would run, would probably cost over £4 billion, or around $7 billion, for 71 kilometres of track.[73] The cost of the 500-kilometre maglev line which might one day cut through the mountains between Tokyo and Osaka has been estimated at $82.5 billion.[74] These prices suggest costs of $40 million to $165 million per kilometre. The French have been able to build TGV lines for much less, however. The Méditerranée link cost about €23 million ($28 million) per kilometre, and the Atlantique €10 million ($12 million).[75] An 8,000-kilometre track, if the price could be kept as low as the Atlantique's, would cost $96 billion to lay.

It would also take a long time. Public inquiries should be held in the countries through which the line might travel. Rights of passage must be negotiated and the land bought, then a massive engineering project undertaken. With a sufficient sense of urgency the construction of a series of trans-continental TGV lines could be completed within the timeframe considered by this book. But should it be done?

You will not be surprised to see me report that there is a problem. Though trains travelling at normal speeds have much lower carbon emissions than aeroplanes, a discussion paper by Professor Roger Kemp of Lancaster University shows that energy consumption rises dramatically at speeds over 200 kmph.[76] Increasing the speed from 225 kmph to 350 kmph, he reveals, almost doubles the amount of fuel they burn. If the trains are powered by electricity, and if that electricity is produced by plants burning fossil fuels, then a journey from London to Edinburgh by a train travelling at 350 kmph, Kemp's figures suggest, would consume the equivalent of 22 litres of fuel for every seat. An Airbus A321 making the same journey uses 20 litres per seat.[77]

Trains, of course, don't produce condensation trails, so the total global warming effect is smaller. But we are still seeking a 90 per cent *carbon* cut. The 350 kmph link between London and Scotland that Tony Blair appeared to endorse in 2004[78] would, by comparison to flying, deliver a 10 per cent carbon rise. Even if speeds were confined to 250 kmph, Kemp's graph suggests, trains would still consume 14 litres of fuel per seat, giving us a carbon cut of just 30 per cent.[79] In reality, the effects of ultra-high-speed trains would be worse than this, for they would draw people not only out of planes, but also out of slower trains and coaches.

Ultra-high-speed trains, in other words, can be part of the solution only if they run on electricity and it is provided by renewable power and fossil fuel combustion with carbon capture and storage. As they are likely to use only a small percentage of a nation's total power, this should be possible, within the limits set by this book. Trains of this speed powered by engines using their own fuel must be ruled out altogether. But if we are to keep using the railways for passenger transport, the cheaper and more environmentally responsible approach is to keep the average speed of our trains to below the current maximum (in the United Kingdom) of roughly 180 kmph. High performance and low consumption are, again, at odds.

The fastest ocean-going passenger ships cruise at about 30 knots, or 54 kmph. They could go faster. A company called BGT Industrial claims to have made the engines for a freighter which can travel at 70 knots (130 kmph).[80] Even if passenger ships could travel this fast, however, it is still just one seventh of the speed of an airliner. And it is not clear that ships offer carbon savings anyway.

It is remarkably hard to obtain comparative figures for fuel consumption, but George Marshall of the Climate Outreach Information Network has conducted a rough initial calculation for the *Queen Elizabeth II*, the luxury liner run by Cunard, which cruises at between

25 and 28 knots (45–50 kmph). It has to be said that the *QEII* does not exactly optimize its space. It contains seven restaurants and seven lounges, a branch of Harrods and dozens of other shops, cabins big enough for dinner parties, and 920 crew members to serve just 1,790 passengers.[81] But even taking all this into account, the figures don't look good. Cunard says the ship burns 433 tonnes of fuel a day, and takes six days to travel from Southampton to New York. If the ship is full, every passenger with a return ticket consumes 2.9 tonnes. A tonne of shipping fuel contains 0.85 tonnes of carbon, which produces 3.1 tonnes of carbon dioxide when it is burnt. Every passenger is responsible for 9.1 tonnes of emissions.[82] Travelling to New York and back on the *QEII*, in other words, uses almost 7.6 times as much carbon as making the same journey by plane.

Short-haul shipping could be even worse. An initial calculation Roger Kemp made for a car ferry to Norway suggests that at 48 kmph the carbon emissions per kilometre are roughly twenty times greater than those produced by a train travelling at 200 kmph and several times greater than a plane's.[83] Again, car ferries are an inefficient means of shifting people, as the vehicles they carry weigh more than the passengers; even so, his estimate gives us further cause to be gloomy about ships. There are some technical measures – such as designing the hull to create air pockets or coating it with slippery polymers,[84] or using a 'towing kite'[85] – which could reduce a ship's emissions; but most are speculative and, it seems, their possible applications are limited. Unless we are prepared to travel very slowly, as much of our freight does, shipping is not the answer.

Becoming rather desperate now, I have looked into airships: craft kept aloft by gases that are lighter than air. In some respects they are quite promising – according to the Tyndall Centre, their total climate impact is 80–90 per cent lower than that of aircraft.[86] (This is not the same as an 80–90 per cent carbon cut, however, as it takes into account the other emissions jet planes produce.) This could be improved with better engines or possibly even hydrogen fuel cells. Kevin Anderson of the Tyndall Centre points out that if they were suspended by means of hydrogen rather than helium, the gas could be drawn out of the ballast as they travel and used for fuel.[87] This is quite a neat proposal. At present, airships become lighter as their fuel is consumed, and therefore harder to control. Anderson's proposal, if workable, could allow them to retain roughly the same buoyancy throughout the trip. Despite the residual public memory of the *Hindenberg* disaster, they appear to be safe. They have a range of up to 10,000 kilometres. But, though faster than ships, their top speeds are currently confined to around 130 kmph: a flight from London to New York would take about 43 hours. They also have trouble landing and taking off in high winds and making way if the wind is against them. This makes both take-off times and journey times less reliable than those of jets. But if we really have to cross the Atlantic, and we are to prioritize the reduction of carbon emissions, airships, surprisingly, might be the best kind of transport.

But now I really have run out of options. Not only is there no means of cutting emissions from planes to anything resembling the necessary level, but there is no form of transport which achieves much more than a quarter of their speed without producing comparable quantities of carbon. There is simply no means of tackling this issue other than to reduce the number, length and speed of the journeys we make.

If we were to overlook the additional climate-changing effects of flying and assume – perhaps optimistically – that a 20 per cent improvement in fuel efficiency is possible by 2030, we would need to cut the number of flights we make by 87 per cent to meet my target. But if we take the other climate impacts into account, and remember that fuel efficiency is likely to be counteracted by vapour formation, we must cut flights by over 96 per cent. If long-range propeller planes took the place of jets, however, and flew below the level at which condensation trails are formed, we might be able to get away with a smaller reduction. The alternative is to cut the carbon emissions produced by other parts of the economy by more than 90 per cent in order to accommodate a greater contribution from flying. To do this, we would have to argue that flying is more important than heating or lighting. As it is practised only by those who are – in global terms – rich, this argument would be difficult to sustain.

Again I feel I should remind you that this is not an outcome I have chosen. If you don't like it, you must find a means of proving me wrong, and it had better be more persuasive than the proposal to transport people by means of cosmic energy that one of my readers sent me.

So I offer you no comfort in this chapter. A 90 per cent cut in carbon emissions means the end of distant foreign holidays, unless you are prepared to take a long time getting there. It means that business meetings must take place over the internet or by means of video conferences. It means that trans-continental journeys must be made by train – and even then not by the fastest trains – or coach. It means that journeys around the world must be reserved for visiting the people you love, and that they will require both slow travel and the saving up of carbon rations. It means the end of shopping trips to New York, parties in Ibiza, second homes in Tuscany and, most painfully for me, political meetings in Porto Alegre – unless you believe that these activities are worth the sacrifice of the biosphere and the lives of the poor. But I urge you to remember that these privations affect a tiny proportion of the world's people. The reason they seem so harsh is that this tiny proportion almost certainly includes you.

Recognizing that it was possible for a human being to fly; then that it was possible for a human being to fly long distances; then that it was possible for many humans to do so; then that it was possible for *you* to do so, required a series of imaginative leaps. It required the construction by the people of the twentieth century of a possible world which did not exist before. No one in Europe ever thought of shopping in New York or visiting friends in Australia before planes allowed them to do so. Recognizing that while it is still possible for a human being to fly, it will no longer be possible for many humans to do so, indeed that it will no longer be possible for *you* to do so, requires a similar series of imaginative efforts. But if it was possible to construct one alternative world, it is surely possible to construct another, and to adjust ourselves to that world (scarcely conceivable as it now seems) just as we adjusted to the other – even less conceivable – existence.

I do not pretend that this will be easy, or that my finding will win me any friends. Those whose freedoms must be curtailed happen to be members of the world's most powerful classes. Worse still, they happen to be us. The promises we have been made – of tropical sunlight in the dead of winter, of one-week safaris in the Maasai Mara, of heroic missions to rescue the bridgeless people of the Peruvian Andes, of the sampling of pleasant fruits and princely delicates throughout the new-found world – have shaped our expectations, the pictures we carry of our future lives. We have come to believe we can do anything. We can do anything. Accepting that we no longer possess the powers of angels or of devils, that the world no longer exists for our delectation, demands that we do something few people in the rich world have done for many years: recognize that progress now depends upon the exercise of fewer opportunities.

Rationing alone will not make all the necessary decisions for us. If airport capacity is permitted to keep expanding, indeed if it is not deliberately reduced, then flying will break the rationing system just as it will, on current projections, break the Emissions Trading Scheme. The gulf between what we could do and what we should do would simply be too great: the political clamour to expand the allocation to permit us to make use of the growing opportunity insuperable. Even before a rationing scheme is in place we must lobby for a moratorium on all new runways. This campaign in many rich nations – including the United Kingdom – has already begun. Climate-change campaigners have joined forces with the people who live close to where the runways might be built, who fear that their lives will be ruined.[88,89]

I have sought the means of proving otherwise, not least because it would make my task of persuading people to adopt the proposals in this book much easier. But it has become plain to me that long-distance travel, high speed and the curtailment of climate change are not compatible. If you fly, you destroy other people's lives.

10

Virtual Shopping

*Believe me, Master Doctor, this makes me wonder above
the rest, that being in the dead time of winter and in the
month of January how you should come by these grapes.*
                                    Doctor Faustus, Act IV Scene II[1]

Reviewing the last nine chapters, I discover that I have been uncharacteristically kind to business. I have concentrated on the measures which might reduce the amount of carbon we emit directly, rather than that released on our behalf by industry. So I will attempt to salvage my reputation by examining two enterprises: retailing and the manufacture of cement. I have chosen them because their carbon emissions, which are very great, are, at first sight, particularly hard to address.

The business practices of the superstores sometimes look like a carefully designed project to destroy the biosphere as swiftly as possible. Their freight transport arrangements, for example, seem almost perversely designed to maximize the distance travelled. Among the examples I came across when researching another book (*Captive State*) was that of the vegetables being sold in two superstores on the outskirts of Evesham in Worcestershire, in central England. They had been grown just two kilometres from the town. First they were trucked to Herefordshire, some 70 kilometres away, then another 130 kilometres or so to a pack-house in Dyfed in south Wales, then a further 290 kilometres to a distribution depot in Manchester, then 180 kilometres back to Evesham.[2]

Like Dr Faustus, who brought grapes from the southern hemisphere 'by means of a swift spirit that I have' to feed a pregnant duchess, the

189

supermarkets respond to – and help to create – a demand for unseasonal produce. Before air freight was cheap, no one but Queen Victoria thought of demanding perishable food from the other side of the world. She is alleged to have offered a large reward to anyone who could bring her a fresh mangosteen. Lacking Faust's powers, no one was able to claim it. Today it is often harder to find a British apple in the shops than a mango or a papaya. Herrings, in the United Kingdom's superstores, are rarer than tiger prawns. But the shops take advantage not only of the other hemisphere's contrary summer, but also of its lower labour costs, cheaper land and economies of scale. We are all now familiar with out-of-season Coxes from New Zealand on our shelves – which taste like Kleenex soaked in Diet Coke – when our own apples, perfectly ripe, are falling from the trees through lack of buyers. We have all now seen the potatoes and onions marked 'South Africa' or 'Chile' or 'Australia', and the bottled water, effectively indistinguishable from that which comes from our taps, imported from the other side of Europe.

But this is not another section on transport. With the introduction of carbon rationing, unnecessary freight of this kind will simply be priced off the shelves. I know that some people will find this hard to swallow. In her book *How to Eat*, the celebrity chef Nigella Lawson dismisses concerns about long-distance transport thus:

If you live in the Tuscan hills, you may find different lovely things to eat every month of the year, but for us it would mean having to subsist half the time on a diet of tubers and cabbage, so why shouldn't we be grateful that we live in the age of jet transport and extensive culinary imports? More smug guff is spoken on this subject than almost anything else.[3]

Lawson's requirement for asparagus in October plainly takes precedence over other people's requirement for survival. But she also betrays a limited imagination. Rocket, lamb's lettuce, purslane, winter cos, land cress, kale, leeks, chicory, pak choi, choi sum, mizuna, komatsuna, mooli, winter savory, coriander, parsley, chervil, spring onions, spinach, sorrel and chard will grow through the winter in the United Kingdom. Some need cold frames or cloches to protect them from the lowest temperatures, but none requires a heated greenhouse. Carrots, parsnips, potatoes of all kinds, beetroot, onions, garlic, swedes, pumpkins, squashes, celeriac, salsify and scorzonera can be stored without refrigeration. Scores of old apple varieties, among them some of the best ever cultivated – Ashmead's Kernel, Ribston Pippin, Aromatic Russet, Belle de Boskoop, Pitmaston Pineapple, Allen's Everlasting, Court Pendu Plat, D'Arcy Spice – have been developed to last through the winter in an insulated shed. Even in Marlowe's day, horticulture was sufficiently advanced to produce an apple – the Winter Greening or Apple John – which would still be edible, if wrinkled, *two years* after it was picked.[4] In *Henry IV Part I*, Falstaff complains that

my skin hangs about me like an old lady's loose gown; I am withered like an old apple-john.[5]

And this is to say nothing of the products (some, like damson jam and raspberry vinegar, quite exquisite) of smoking, salting, drying, pickling and preserving. If Nigella Lawson can't summon the wit to make a decent meal from ingredients like this, she should find herself another job. In short, given that much of the food brought here from afar is picked so early that it rots before it ripens, it is not hard to see how our diet, even for those of the most refined and demanding tastes, could in fact be improved by means of geographical restriction.

But even if we were (because the solution to this profligate use of fuel is so obvious) to disregard the means by which our retailers obtain and distribute their goods, their consumption of energy remains astounding. The table below gives the figures for shops and other buildings. Given that nothing except money is made in most

sector	space heating and domestic hot water (kWh per m^2)	electricity (kWh per m^2)
warehouses	64	81
local government offices	95	39
commercial offices	147	95
factories	245	47
retail	185	275

Source: Royal Commission on Environmental Pollution.[6]

shops, the fact that their heat demand is not far behind that of factories, while their requirement for electricity is almost six times as great is, at first, scarcely credible. But then you think your way around. As you come through the door of a supermarket, a unit above your head blasts you with hot air in the winter and cold air in summer (sometimes, when the manager has not been paying attention, it is the other way around). You must stand blinking for a moment as your eyes adjust to the lights. Then you walk past banks of fridges and freezers which have *no doors*. This would be impossible to believe, if it were not by now one of the most ordinary facts of life. But, though you walk through valleys of ice, you remain warm. All day long, the freezers and the heaters must fight each other. They must do so in a building which is huge, generally uninsulated and often widely glazed: that is capable, in other words, of trapping neither heat nor cold.

In the hope of finding the means by which the superstores might achieve major cuts in energy consumption, I visited a senior manager from one of the big chains. I cannot name him or his company, but its practices appear to be no worse than those of its competitors. He began by giving me an idea of where the energy goes.

The heaters over the doors each have a rating of 50 kilowatts. This is roughly seventeen times as powerful as a standard domestic fan heater. The aisles are lit to an intensity of 1,000 lux, which is about the same saturation as a TV studio, and two or three times that of an office. The counters are brightened with spotlights – at up to 2,000 lux. Fish, in particular, must sparkle, so they are lit with ceramic discharge metal halide lamps, which are otherwise used to illuminate castles and cathedrals at night. You begin to understand what this means when you remember that fish have to be kept on ice, while lamps of this brightness could fry them. But, my contact told me, 'If you light it, it will sell. You can't afford not to do it.' Between 20 and 25 per cent of his chain's energy budget, he told me, is spent on lighting.

Most of the rest – 64 per cent – is used for refrigeration. Every open freezer costs his firm £15,000 a year. When the company fitted glass doors to the top half of its cabinets (the vertical portion, at eye level), it cut its refrigeration budget by around one quarter. It could do the same again if it fitted doors to the chest freezers underneath,

'But the traders [the managers of the stores] won't have it.' When customers open a freezer door then close it, it can steam up, obscuring the view.

We then discussed what his company might be able to do to reduce its energy demand. He appeared to have looked into almost every possibility, and the answer, he had found, was not very much. The ceramic discharge spotlights the chain is now using to light its counters are three times as efficient as halogen lamps, and in a few years' time it might be possible to replace them with light-emitting diodes, which are three times better still. At present, however, LEDs 'just don't deliver the punch we want', and the demand for lighting rises inexorably. Similarly, though the company has started replacing its older models,

We can't reduce the overall refrigeration budget, as the number of fridges is growing faster than our efficiency measures.

If the managers took greater care in turning off appliances when they are not needed, they might be able to cut the energy budget by 5 per cent. Ventilating a store at night would reduce the amount of air conditioning it needs when the customers arrive, but natural ventilation, he said, is unpopular with the traders, as it is hard to guarantee consistent temperatures. It might be possible, in new stores, to pump some of the excess heat into the ground during the summer, and tap it in the winter. I suggested that his chain could gather heat from pipes laid under its car parks, but he was unenthusiastic.

Instinctively, I would say it's a no-goer. During construction, there are trucks and bulldozers running all over the site.

They would smash the pipes before the tarmac was laid. He had looked into 'sunpipes': tubes which bounce natural light into the store.

They are promising, but quite expensive, and they don't reduce our capital costs, as we still need full lighting when it gets dark.

Local wind power, he said, is useless: 'on a cold, still day, you're stuck.' Solar panels are far too expensive and – in the buildings he had visited – produced much less electricity than the installers claimed.

Combined heat and power currently costs more than his bosses are prepared to pay.

All that counts is cost . . . Everyone has to move, or no one moves. If we do it and nobody else does, we're lost. Who is going to be the first to do it? . . . Retail is a very harsh environment.

One chain – J. Sainsbury – claims it has been the first to blink. In 1999 it opened 'the UK's most environmentally responsible super-market' on the Greenwich peninsula in London.[7] It uses an earth cooling system for air conditioning, natural light from north-facing windows, a gas-fired combined heat and power station, solar panels and two wind turbines. These innovations, the company claimed, were 'expected to reduce energy consumption by up to as much as 50 per cent compared to a standard store of a similar size and operation'.[8]

Savings of this magnitude, if replicated everywhere and combined with measures to reduce the carbon content of our electricity and heat that I discussed in Chapters 5 to 7, could deliver a total cut of 90 per cent, assuming that the supermarkets' energy use does not continue to rise. But are they real? My researcher contacted J. Sains-bury three times, hoping to obtain the operational figures, and to discover whether or not they had been independently audited.[9] Six months later, we are still waiting for a response. Until it comes, jaded by the false claims of other companies, I reserve the right not to believe a word it says. Indeed, the only firm figures I can find for this 'watershed in supermarket architecture'[10] give me further cause for suspicion. The store's two wind turbines, which Sainsbury's cus-tomers see when they enter the car park, are each 3.6 metres in diameter.[11] This suggests, at mean windspeeds of 4 metres per second, that their average combined output is a little over 0.4 kilowatt hours* – a microscopic fraction of the power the store must use. Even this is likely to be generous, as they stand just 12 metres from the ground, and their poles support advertising hoardings, which must create turbulence.

I have already uttered the dread words 'car park', which hint at

---

* *Building for a Future* gives the output of a 3.5-metre turbine at 1766 kilowatt hours per year at 4.0 metres per second.[12]

the third and perhaps least tractable impact of the superstores. Car journeys account for 62 per cent of the visits made to shops,[13] and for almost all the trips to shops outside the town centres. The Department for Transport's figures suggest that shopping accounts for 20 per cent of the journeys made by the United Kingdom's drivers, and 12 per cent of the distance covered.[14]

It is not easy to see how people visiting out-of-town shops could be persuaded to travel by other means: carrying a week's worth of groceries on the bus or by bicycle offers little by way of entertainment. To judge by the size of the car parks they keep building, and their efforts – so far successful – to prevent the government from imposing parking fees on their customers,[15,16] the superstores have no intention of encouraging people to change their habits. Their customers show no inclination to be encouraged. In this respect at least, shops appear to be locked into profligate levels of energy use.

In other words, if the existing infrastructure and existing shopping patterns are to persist, the scope for cutting emissions is limited. But it seems to me that there might be a means of reconciling a 90 per cent carbon cut with quick and easy shopping. My proposal could solve both the transport problem and the refrigeration and lighting problem in one stroke. It's hardly new: it is a revolutionary idea called delivery.

According to the Department for Transport,

a number of modelling exercises and other surveys suggest that the substitution of private cars by delivery vehicles could reduce traffic by 70 per cent or more.[17]

Every van the superstores dispatch, in other words, appears to take three cars off the road. Already the supermarkets (as well as plenty of new companies) are, with the help of television and the internet, reverting to a way of doing business they abandoned decades ago. At least a couple of times a week, I see a Tesco delivery van coming down my street. I suspect that I am fated to be appropriately run over one day by an organic vegetable box van, as so many of them now infest my home town. The Office of National Statistics calculates that deliveries comprise about 4 per cent of total retail sales; the transport department's paper suggests this might be an underestimate.[18] Sales of goods on the internet appear to be growing almost exponentially, though for obvious reasons there are no reliable figures.

None of this, by itself, will save us. Most shopping will continue, under the existing system, to be done by car, and the stores will continue to freeze their goods and heat their customers in the open air. But if deliveries were to replace shopping in distant stores *in its entirety*, we might have a formula for an 80 or 90 per cent reduction in carbon emissions, even before we consider the use of renewable power and carbon capture and storage.

My proposal amounts to this: that the out-of-town stores are gradually replaced with warehouses. No one visits them but suppliers and the company's staff. They require no supersaturated lighting, no open freezers, no heaters above the door.

Warehouses, as you can see from the table printed a few pages ago, use – per square metre – about 35 per cent as much heat and 29 per cent as much electricity as shops. But you can pack many more goods on to a square metre of warehouse than you can on to a square metre of superstore. There is no need for displays, broad aisles or cash tills, and the shelves can be built much higher. My supermarket contact told me that the stockrooms attached to his stores – through which all their goods must pass, if only fleetingly – are probably responsible for around 5 per cent of the chain's total energy bill.

If, in other words, instead of picking goods from the shelves of their shops – as the superstores do now – then loading them into vans, they were to deliver them straight from the warehouse, not only would they cut the transport emissions caused by collection by 70 per cent, but they could also reduce their static energy consumption* by some 95 per cent. Major shopping trips would, in other words, be eliminated. Local shops (which are much less dependent on cars) could remain open, but they would have to start introducing the kind of efficiency measures which apply to the rest of the economy.

It seems to me that there may be a further environmental advantage to this proposal: that fancy packaging would no longer be considered necessary. As the point of sale is the computer or the television or the telephone, and the visual stimulus is the catalogue (either electronic

---

* By this I mean the energy used inside the store, rather than on the road.

or printed), the goods can be delivered in plain parcels, using no more paper or plastic than is required to keep them clean and fresh. This, for example, is how I buy my vegetable seeds. The companies I use publish online catalogues full of beautiful pictures of healthy plants (which look nothing like the slug-reamed specimens I grow) then send me the seeds in brown paper envelopes. This presentation does nothing to diminish either the excitement I feel when they arrive or the eventual disappointment.

At first glance, the business case for a complete transition to virtual shopping seems as sound as the environmental case. Capital, staff and overheads costs are all reduced. But it is naïve to imagine that this model would recommend itself to the existing stores. Their profitability depends in part upon the scarcity of sites for which planning permission can be obtained. In many places, the local market has been monopolized by a single company. Because costs are lower and available sites more numerous, the warehouse model allows far more companies to play, as the explosive growth of small organic box schemes shows.

But whether they like it or not, the superstores have already demonstrated that it can be done. Their deliveries are more reliable than those of other companies. While many firms keep you waiting indoors all day, the market leader in internet shopping – Tesco – promises to deliver within a two-hour period.[19] In most parts of the country, you can choose to receive your groceries at any time between 9 in the morning and 11 at night during weekdays, and during most daylight hours over the weekend.[20] In other words, whatever the firms might think of it, this kind of shopping is likely to be more convenient for their customers than the existing model – which is why the sector seems to be growing so quickly. If my proposal were adopted, it could also be a good deal cheaper. Some people object that deliveries which depend on telecommunications exclude those without the technology; but far fewer households are without televisions or telephones than without cars: my proposal is, in other words, more inclusive than the current system.

But the closure of the out-of-town stores, for the reason I have given, is unlikely to happen by itself. It would need to be stimulated either indirectly – by our rationing scheme – or directly, by regulation.

In either case, however, it seems to me that this switch involves no significant reduction in human freedom, unless the right to travel round the ring road to an overlit glass box and stand in a queue is to be represented as an unalienable component of life, liberty and the pursuit of happiness.

I have learnt from bitter experience that it is not easy to interest people in cement. But it interests me, partly because its carbon emissions are not confined to those produced by burning fossil fuel. Making 'Ordinary Portland Cement', which is the grey stuff known to almost everyone as plain 'cement',* is a matter of turning limestone (calcium carbonate) into calcium oxide. This means producing carbon dioxide.†

The chemical process – 'calcination' – releases around 500 kilograms of the gas for every metric tonne of cement it makes.[22] The raw materials must also be ground and then heated to about 1450°. Altogether, according to a study published in the *Annual Review of Energy and Environment*, the manufacture of 1000 kilograms of cement emits, on average, 814 kilograms of carbon dioxide.‡[23] This does not take into account the energy costs of quarrying and transport. It is probably fair to say that a tonne of cement produces about a tonne of carbon dioxide.

According to David Ireland of the Empty Homes Agency, writing in the *Guardian*, a house requires, on average, 25 tonnes of concrete for the foundations and floors, and 4 tonnes for mortar and rendering. One tonne, he says, is wasted.[24] He suggests that this results in the release of 30 tonnes of carbon dioxide. But he appears to have confused concrete with cement. The concrete for foundations and floors contains roughly one part of cement to five of sand and gravel. Even so, this suggests, if his other figures are correct, that every new home requires about 5 tonnes of cement, which emits 5 tonnes of carbon dioxide. Even if we were to forget about all the other materials from

---

* Its promoters, in the early nineteenth century, claimed it resembled the fine freshwater limestones quarried on Portland Bill in Dorset, southern England.
† In simplified form, $CaCO_3 > CaO + CO_2$. In reality, Ordinary Portland Cement contains silicates: $5CaCO_3 + 2SiO_2 > (3CaO, SiO_2) (2CaO, SiO_2) + 5CO_2$.[21]
‡ 222kg C × 3.667.

which a house is built, this equates to four times a single person's annual carbon ration in 2030.

Altogether, depending on whose figures you believe, cement produces between 5 and 10 per cent of the world's manmade carbon dioxide.[25,26,27] In the United Kingdom, where we are replacing our houses very slowly, it accounts for just under 2 per cent.[28] Partly because of the construction booms in South and East Asia, global cement production is growing by about 5 per cent a year.[29]

It appears to be as easy to capture the carbon dioxide produced by both the calcination process and the combustion of the fuel which fires the kilns as it is to capture the gas from power station exhausts.[30] Burying it, however, is another matter. Unlike power stations, cement works are geologically constrained: they must be built beside a source of suitable limestone. Though all are to be found in sedimentary basins, there is no evident relationship between limestone outcrops and the salt aquifers and gas and oil fields in which carbon dioxide might be buried. Although most cement plants in the United Kingdom probably lie within 500 kilometres of a suitable aquifer (which is roughly the distance over which the gas can be pumped economically[31]), until maps of appropriate burial sites are published, I can't tell whether or not the technology could be universally applied. But I can state with confidence that the United Kingdom – which is the most geologically diverse region of its size on earth – is atypical. In larger countries, where the geology is less varied, there must be cement plants whose emissions cannot be stored below ground.

Only part of the problem can be solved by building less. There are plenty of uses of concrete – such as new motorways and runways – which are unnecessary and unsustainable. The environmental cost of cement manufacture provides a further powerful argument to stop expanding the transport networks. But, as I mentioned in Chapter 4, many of the homes in the United Kingdom are simply incapable of keeping heat in and weather out. If it can be demonstrated that there are major and rapid carbon savings to be made by knocking them down and rebuilding them, the rate of demolition should increase. The passive houses with which I would like to replace them use more materials than ordinary homes of the same size, as they require a high 'thermal mass'. Even if this were not the case, they couldn't be built entirely from the remains of the old homes: concrete can be recycled, but not as cement – it is used as 'aggregate', or crushed stone.* We'll also need plenty of concrete in which to set our wind turbines.

So the next obvious step is to find means of making the cement we use go further. At first sight, something called AirCrete (or 'autoclaved aerated concrete') seems to provide the answer.[33] When cement is mixed with quicklime, sand, water and aluminium powder, it rises in the mould like a loaf of bread. AirCrete blocks are strong and they keep the heat in. Because between 60 and 85 per cent of their volume is air,[34] they contain much less cement than solid blocks. Unfortunately, the carbon savings appear to be thrown into reverse by the aluminium powder.[35] This is the component that acts as the yeast in the concrete dough, generating the bubbles which make it rise. Though much smaller quantities are used, the energy costs of smelting aluminium are around forty times as great as the energy costs of manufacturing cement.[36]

A more promising approach is the replacement of ordinary mixtures with High Strength Concrete, which contains additives such as silica fume† and finely ground fly ash. Because it is extremely strong,‡ it allows builders to halve the weight of the materials they use.[38] The additives are quite expensive, but because the volume of materials and the transport costs are smaller, the overall cost is generally lower.[39]

The problem is that the industries which produce silica fume are moving out of the rich nations, which means that the carbon costs of transport will rise, while fly ash is, on the whole, the product of burning coal in power stations, which I am seeking to prevent. A similar problem affects another proposal for reducing the carbon content of cement: mixing it with slag from blast furnaces. Not only are steel-makers closing down in many of the rich nations, but those which remain are switching to newer steel-making technologies (such as the electric arc furnace), whose slags don't have the right properties.[40]

In 1997, a proposal by a materials technologist from Nevada called

---

* According to the Intergovernmental Panel on Climate Change, 'closed-loop cement recycling is not yet technically possible.'[32]
† Silica fume is a waste product of the ferrochromium and ferromanganese industries.
‡ It can carry loads of 6,000 pounds or more per square inch.[37]

Roger Jones caused great excitement in those parts of the media (which I admit are not numerous) that follow developments in the cement industry.[41,42] He suggested turning cement back into limestone, by forcing it to reabsorb the carbon dioxide it loses during the calcination process.

This happens anyway, if very slowly. According to *New Scientist*, 'a large slab of concrete could take 30,000 years to carbonate fully.'[43] This is a little beyond the timescale considered by this book, though at current demolition rates in the United Kingdom, we can expect some of our houses to turn to limestone before they are pulled down. Jones discovered that supercritical carbon dioxide (heated and compressed to just over 1,000 pounds per square inch[44]) can pass straight through concrete, carbonating it in minutes and doubling its strength.[45] The cement would reabsorb all the carbon dioxide it lost during calcination. The supercritical gas could, Jones claimed, simply be sprayed on.[46] Instead of being buried, carbon dioxide captured in cementworks could be poured back into the cement.

In April 2006, I phoned Roger Jones in Nevada to find out what had become of his proposal. 'It really wasn't for concrete, George,' he told me. 'It was for plastics. It's a very expensive process, so we never intended that it should go in that direction.'[47]

This surprised me, not least because a press release still to be found on his company's website claims that his process

transforms common portland or lime cemented materials and clays by treatment with carbon dioxide ... 'Like living coral, now we can take carbon dioxide out of the environment and build our houses with it.'[48]

This, in other words, is a further demonstration of the need to be wary about the speculative claims made by people with a commercial interest. The same consideration makes me cautious about the plans announced by a company called TecEco to produce cement made from magnesium – rather than calcium – oxide, even though in some respects it sounds quite promising.

Magnesium carbonate needs be heated to only 650° to make cement,[49] which greatly reduces the energy costs of manufacturing it. The oxide also seems to absorb carbon much faster than calcium cements do: the man who runs the company claims it takes just a few months and that it ends up stronger than ordinary cement.[50] The problem is that magnesium carbonate is rarer than limestone. This means that it is more expensive and – unless the cement is sold only in the regions which produce it – incurs greater transport costs. But it might be useful in some places.

There does, however, seem to be a solution, and once more it is, in concept, a very old one. It is a material similar to the *pozzolan* cements with which the Romans built the domed roof of the Pantheon and hundreds of other structures, some of which are still standing today.[51] The institute which promotes it claims that it sets so quickly and to so great a strength that 'a heavy Boeing or Airbus can land on a runway' patched with this material only four hours after the repair took place.[52]

While this is not the example I would have chosen, it does testify to properties that no paper I have been able to find disputes. This material sets quickly, appears to be stronger than ordinary cement, lasts longer, shrinks less and is more resistant to fire.[53] 'Geopolymeric cements', as these materials are known, can be manufactured from several kinds of clay and industrial waste and quite a few common sedimentary rocks.* They are cheap and, most importantly, their fabrication produces between 80 and 90 per cent less carbon dioxide than Portland cement.[54,55] This is because they are formed at lower temperatures (about 750°) and the chemical process doesn't depend on releasing carbon dioxide. According to the Australian government's research body, CSIRO, they can be used for every major purpose for which ordinary cement is bought today.[56] The reason why they are not yet widely manufactured is that artificial geopolymers – which don't depend on the rare deposits of natural *pozzolana* (a type of volcanic ash found around the Bay of Naples) used by the Romans – were invented only in the late 1970s.[57] The construction industry is notoriously conservative, and the cement companies have a powerful financial incentive to maintain their existing plants, rather than to start up somewhere else with a different process.

But the answer, or most of it at any rate, appears to exist. By 2030 there should be no plants in the rich nations which are not either

---

* The key components are compounds of aluminium and silica.

burying their carbon or producing geopolymers instead of Portland cement. And if extra materials are needed for building better houses, that growth should be offset, perhaps even reversed, by halting the construction of new roads and runways.

Re-reading this section, I am forced to admit that cement is in fact a rather dull subject. I beg your forbearance on the grounds that to address only those sources of carbon dioxide which are interesting would be to succumb to another kind of aesthetic fallacy.

So I think I might have got there, more or less. I hope I have been able to show that we could cut our carbon emissions by around 90 per cent in all but one of the sectors I have investigated: in the home, on the roads, and in two industries which at first sight seemed particularly difficult to reform. The sector in which I have failed – and I hope I am not letting myself off too lightly here – happens to be the one which is least necessary to our survival. Unlike heating, lighting, travelling to work, building or shopping, aviation is not required to sustain civilization, though its loss from the lives of most of the people who use it today will be keenly felt.

I can't pretend that my proposals are anything other than extremely challenging. They can be implemented only if tackling climate change becomes the primary political effort not just in our own country but in all rich nations. They require a good deal of money and a great deal of political will and expertise to enact. But what I hope I have demonstrated is that it is possible to save the biosphere. If it is possible, it is hard to think of a reason why it should not be attempted. It is true that this effort will disrupt our lives. But it will cause less disruption than the alternative, which is to allow manmade global warming to proceed unhindered.

11

Apocalypse Postponed

O, might I see hell and return again, how happy were I then!
Doctor Faustus, Act II, Scene III[1]

One week before this book was meant to be finished, my daughter was born. I typed Chapter 10 one-handed while she sat on my thigh. As my mind switched from cement to vomit and wind speeds to stomach gas and back again, everything I had been thinking about became – for the first time – real to me.

The position of a writer on subjects like this is an odd one. In order to achieve some grasp of the complex matters about which you care, you must withdraw from the world and enter a shadow land of graphs and tables, equations and projections. In doing so, you must cease to care. The ecosystems you consider become 'carbon sinks' or 'carbon sources', the people become data.

I have spent around two years in Africa, and of the nations I have visited there, I loved Ethiopia the most. Wherever you go in that country, you can make friends almost instantly. Perhaps because people there have long lived so close to the brink of destruction, the conversation often turns to matters of profundity. In the tiny market town of Jinka, in the south of Ethiopia, where most people are illiterate, my brain was worked as hard by the people I met as it has ever been by the inhabitants of my home city, Oxford.*

---

* I sometimes wonder whether the comparative scarcity of death in recent centuries is responsible for the fact that British writers can no longer discuss existential matters with the confidence of Shakespeare or Marlowe.

But I now realize that when writing in this book about 'the Ethiopians' and the impacts of climate change they might suffer, I did not think once about the people I met there. 'The Ethiopians' had become an abstraction or worse: a tool, perhaps, which I could use to help shape my argument. A certain degree of warming might, through the reduction in a certain number of bushels of grain, eliminate a certain number of units of the species *Homo sapiens*. I made no connection between those units and the humans I met there, perhaps because these lively, funny, stimulating people bore no resemblance to a column of figures on a page.

Similarly, when considering what might happen to people in my own country or in other parts of the rich world – in which the human impacts of global warming will be delayed both by our more forgiving climate and by the money we can spend on our protection – I have found the likely effects easy to catalogue but almost impossible to imagine. I can understand, intellectually, that 'life' in this country might not be the same in thirty years' time as it is today; that if climate change goes unchecked it could in fact be profoundly and catastrophically different. But somehow I have been unable to turn this knowledge into a recognition that my own life will alter. Like everyone who has been insulated from death, I have projected the future as repeated instances of the present. The world might change, but I will not.

Underlying this denial is the dissonance with which we face all possible catastrophes: plagues, wars, famines, even death itself. I might be deeply afraid of the impending disaster, but I am also confident that – through the grace of God or the other sources of good fortune that have preserved me so far – it will not apply to me. If part of us did not believe that we could – despite the evidence of other people's mortality – somehow cheat death, we would scarcely struggle to prolong our lives. For, as Doctor Faustus discovered at the end of Marlowe's play, twenty-four years – even 24,000 – is indistinguishable from twenty-four minutes when weighed against eternity.

For the writer, this self-delusion is particularly tempting, for at the back of your mind rests the hope that even if your body dies, your words might live on. Indeed, this appears to lead to the perverse consequence that writers sometimes seem more careless with their lives than other people of their class. What does it matter if you drink yourself to death, when your soul persists 'in memories draped by the beneficent spider'? Indeed there is a certain arrogance, which I have recognized in myself as well as in others, which treats death as a fear belonging to lesser mortals: a terrestrial business which should not trouble those seeking to fashion a gossamer link in that ethereal chain of thought which runs from *Gilgamesh* to the inscrutable future.

But this baby, this strange little creature, closer to the ecosystem than a fully grown human being, part pixie, part frog, part small furry animal, now sixteen days old and curled up on my lap like a bean waiting to sprout, changes everything. I am no longer writing about what might happen to 'people' in this country in thirty years' time. I am writing about her. As she trembles on the threshold of life, the evidence of her mortality is undeniable. It seems far more real than mine. The world I described in Chapter 1, in which unrestrained climate change threatens the conditions which make human life possible, is the world into which she might grow. Global warming is no longer a generalized phenomenon, its victims no longer abstractions. Among them might be my child. Or yours. Or you. Or even me. Of all the complex matters encapsulated in this subject, this, until now, has been the hardest to grasp.

I see too – confronted as I am by biological reality – that even while considering the direst predictions and navigating the granite straits of thermodynamics, I have somehow also entertained a chiliastic belief in salvation. At the back of my mind, at the back, I think, of the mind of everyone who has considered these matters, is the notion that, however real our predicament and the difficulties of escaping from it seem, they cannot possibly be true. Someone or something will save us. A faith in miracles grades seamlessly into excuses for inaction.

The first of these is the hope that many people place – that I sometimes catch myself placing – in unproven technologies. Surely 'they' – the unidentifiable, omnipotent scientists who have taken the place of God and lurk always on the fringes of our consciousness – won't let the collapse of the biosphere happen. Within the necessary timeframe, indeed, so our imaginations tell us, in the nick of time, they will deliver us from evil by inventing a device which harnesses nuclear fusion, artificial photosynthesis, 'hydrinos' or solar power on the moon.*

Every few weeks someone contacts me with a proposal for what is, in effect, a perpetual motion machine. He (it is always a he) can demonstrate to my satisfaction that, unlike all the quacks and cranks and mountebanks I have heard about, he really has solved the problem. He has a special catalyst, or a new equation, or a hotline to God, which demonstrates what all other physicists think impossible: that energy can be created. It sometimes makes me imagine that I have been transported back to Marlowe's day. My only defence against these people is to ask them for an article in a peer-reviewed journal, whereupon I never hear from them again.

This is, perhaps, a little unfair. It could be that someone out there really has developed a new kind of fuel, whose existence defies the recognized predictions of science and that because it defies these predictions no self-respecting journal will dare to publish. All professions are conservative, even those which seek to negotiate the future. But accepting that there are more things in heaven and earth than are dreamt of in our physics labs is a very different matter from relying on a scientific miracle for deliverance. For all the agency this faith affords us, we might as well perform a climate-cooling dance.

Nor is it fair to suggest that all speculative technologies are equally unlikely. While energy from nuclear fusion has always been thirty-five years away, we cannot be sure that it always will be: indeed, every so often, a 'breakthrough' is announced which suggests for a moment that it could be only twenty years away (but this figure also seems to be a constant). In any case, even with the kind of accelerated programme I advocated in Chapter 5, it is hard to believe that a new form of energy, however promising, can be identified, harnessed and then adapted to all the uses to which we would need to put it – heat, electricity, land transport, flight – and then universally applied within the next twenty-four years. To succumb to hope of this nature is as dangerous as to succumb to despair.

The second miracle of deliverance, or excuse for inaction, is related

---

* It might sound as if I am making this last one up, but it is a serious proposal put forward in 2001 at the 18th Congress of the World Energy Council.[2]

to the first one: a belief that a new technology will permit us either to remove carbon dioxide from the atmosphere once it has been released, or to cool the planet by artificial means. One or two such schemes have been attempted at an experimental level. The best known of these is the scattering of iron particles on the surface of the ocean in order to stimulate the growth of plant plankton. The idea is that the plankton, as they multiply, absorb carbon dioxide from the surface waters, then sink with their burden to the depths of the ocean, removing the gas for all time. It doesn't work. Modellers at Princeton have shown that hardly any of the gas the plankton absorb is removed from the surface of the sea.[3] At the same time, because it mops up oxygen, fertilization by iron stimulates the production of methane.[4] It seems likely that this technique, as well as wrecking the ecology of the oceans, would cause more global warming than it cured.

Other people have advocated using chemical scrubbers – like those which would remove the carbon dioxide from the exhausts of power stations – to extract the gas from the open air. Perhaps unsurprisingly, none of the accounts I have read have costs attached to them.[5,6,7] I phoned the company promoting the idea, Global Research Technologies, in Tucson, Arizona, hoping to obtain figures for the amount of money and energy required to extract a tonne of carbon dioxide from the atmosphere. After discussing my request, they called me back to say 'the answer is "no comment".'[8] My guess is that the costs would be astronomical.

John Latham, at the National Center for Atmospheric Research in Boulder, Colorado, has been experimenting with devices which spray seawater into the air. They will, he hopes, create clouds which would screen out some of the sunlight reaching the earth.[9] This project could be exceedingly dangerous: it seems that small salt particles, while generating mists, could actually retard the development of rain-bearing clouds, causing droughts in the countries downwind.[10]

When he died, Edward Teller, the man who developed the hydrogen bomb, left us a second generous legacy: the idea of flooding the atmosphere with particles of metal or other materials, which could reflect light of certain wavelengths. His disciples (who also work on nuclear weapons by day) have developed this idea, proposing to launch a million tonnes of tiny aluminium balloons, filled with hydrogen, every year. It is hard to decide which of their activities is more dangerous; their scheme, if implemented, is likely to eliminate the ozone layer.[11] Experimenting with mass death appears, for some people, to have become a habit.

The third messiah, paradoxically, is peak oil. I have lost count of the number of people who have explained to me that we don't need to worry about climate change, because before it advances too far, global oil supplies will go into decline, the price will rise exponentially and the motorists and airline passengers will be forced to stay at home.

Now it could well be true that an oil peak, if it takes place before proper contingency measures have been developed, could throw the world into a depression so catastrophic that it effectively brings industrial civilization – and therefore its carbon emissions – to a halt. I am not among those who welcome this prospect. Nor do I believe we have a firm idea of when it is going to happen. When I first came across predictions that oil supplies could peak very soon – one geophysicist announced in 2003 that he was '99 per cent confident' it would happen in 2004[12] – I found them persuasive. But the more I read, the less certain I become. As in other such cases, you can find people and data on both sides of the debate with an equal claim to be taken seriously.[13,14,15,16] It could well be true that petroleum will peak within the next 10 years; it could also be true that it will take 30. If this is the case and if we have placed our faith in the decline of oil supplies while simultaneously failing to do anything to prevent it (not, I am afraid to say, an unlikely proposition), we could find ourselves facing catastrophic climate change *and* an unprecedented global depression.

In one respect, peak oil could exacerbate climate change. The report commissioned by the US Department of Energy, which I mentioned in Chapter 3, advocates anticipating the problem by greatly increasing the production of synthetic fuel – artificial crude oil – from oil sands and coal.[17] Needless to say, synthetic fuels cause even more climate change than petroleum.

The fourth excuse for inaction is more mundane. This is the idea that we can keep buying our way out of trouble. In his book *The Rise of the Dutch Republic*, published in 1855, John Lothrop Motley describes the means by which the people of the Netherlands in the fifteenth and sixteenth centuries could redeem their sins.

> The sale of absolutions was the source of large fortunes to the priests ... God's pardon for crimes already committed, or about to be committed, was advertised according to a graduated tariff. Thus, poisoning, for example, was absolved for eleven ducats, six livres tournois. Absolution for incest was afforded at thirty-six livres, three ducats. Perjury came to seven livres and three carlines. Pardon for murder, if not by poison, was cheaper. Even a parricide could buy forgiveness at God's tribunal at one ducat, four livres, eight carlines.[18]

Today you can find the tariffs for crimes about to be committed on noticeboards erected throughout cyberspace. 'Carbon offset' companies promise to redeem the environmental cost of your carbon emissions by means of intercession with the atmosphere: planting trees, funding renewable energy projects in distant nations and doubtless, somewhere, helping Andean villagers to build bridges. Just as in the fifteenth and sixteenth centuries you could sleep with your sister, kill and lie without fear of eternal damnation, today you can leave your windows open while the heating is on, drive and fly without endangering the climate, as long as you give your ducats to one of the companies selling indulgences. There is even a provision of the Kyoto Protocol permitting nations to increase their official production of pollutants by paying for carbon-cutting projects in other countries.* I will not attempt to catalogue the land seizures, conflicts with local people, double counting and downright fraud that has attended some of these schemes. That has been done elsewhere.[19,20,21] My objections are more general.

The first is that accurate accountancy for many carbon-offset projects, however honest the attempt, is simply impossible. You can determine, for example, that a flight to New York results in the production of a certain weight of carbon emissions, and you can work out how much carbon a particular tree of a particular species and particular size contains. You can then divide the tonnage of carbon from your flight by the tonnage of carbon contained in one tree, and

---

* The Clean Development Mechanism.

work out how many you would need to plant in order to recoup your emissions. The result will bear no relation to reality.

Planting trees, for example, means not planting – or not leaving – something else on the same land. You have no means of knowing what, in twenty years' time, might have stood in their place. If the answer is other trees, then to determine the real carbon uptake caused by your actions, you would have to subtract the carbon that might have been from the carbon that is. As you have no means of determining the value of the first figure, you have no means of completing the sum.

Planting trees in one place might kill trees elsewhere, as they could dry up a river which was feeding a forest downstream.[22] By taking up land which might otherwise have been used to grow crops, it could drive local people to fell forests elsewhere in order to feed themselves. Your trees might die before they reach maturity, especially as their growing conditions change with global warming. Timber poachers could fell them; a forest fire could fry them. In other words, in flying to New York we can be sure that carbon dioxide has been released. In paying to plant trees, we cannot be sure that it will be absorbed.

More importantly, a tonne of carbon saved today is far more valuable, in terms of preventing climate change, than a tonne of carbon saved in twenty years' time, for the reasons I discussed in Chapter 1. But, as if to show how little they care about real impacts, none of the offset companies I have come across uses discount rates for its carbon savings (which would reflect the difference in value between the present and the future). I think it is fair, therefore, to accuse them all of systemic false accounting, even if it is not intentional.

The United Kingdom's Forestry Commission notes that

the rate of carbon accumulation is relatively low during the [trees'] establishment phase (and may even be negative as a result of carbon loss from vegetation and soil associated with ground preparation). This is followed by the full-vigour phase, a period of relatively rapid uptake, which levels off as the stand reaches the mature phase, and then falls.[23]

Trees can take sixty years or more to reach maturity.

Even the projects which promise to retrieve our carbon emissions by giving people in poorer nations better stoves or energy-efficient lightbulbs take time to work, as they rely on the difference over the years between the carbon which would have been generated by using the old models and the carbon the new ones produce. At best these schemes merely delay the point at which emissions are saved. At worst, they allow us to believe that we can carry on polluting, just as, before the Reformation, the sale of absolutions encouraged people to believe that they could carry on sinning. I cannot think of a more effective means of postponing the hard choices we need to make now.

But my main objection is this: that in order to deliver a carbon cut of the size I have discussed, *everyone* will have to limit their emissions, either today or, in the poorer nations, in the future. There is no choice to be made about whether to abstain from flying or to help poorer people buy better lightbulbs. We must abstain from flying *and* help poorer people buy better lightbulbs. Buying and selling carbon offsets is like pushing the food around on your plate to create the impression that you have eaten it.

I have sought to demonstrate that the necessary reduction in carbon emissions is – if difficult – technically and economically possible. I have not demonstrated that it is politically possible. There is a reason for this. It is not up to me to do so. It is up to you.

Those of us who are already campaigning to reduce the impact of climate change cannot do it by ourselves. Given that this is the greatest danger the world now faces, we are astonishingly few. It appears to be easier to persuade people to protest against the termination of a favourite theme tune, or the loss of imperial measures, or, for that matter, against speed cameras and high fuel prices, than to confront a threat to our existence. There is an obvious reason for this: in those cases something is being done to us. In this case we are doing it to ourselves. In fighting climate change, we must fight not only the oil companies, the airlines and the governments of the rich world; we must also fight ourselves.

The problem is not that no meaningful progress has been made at the international climate talks. The problem is that we have not wanted it to happen. It is true that the governments of the United

States and Australia have done everything in their power to prevent the talks from succeeding or even from taking place. It is true that the defining feature of these negotiations is that someone else is always to blame. The governments of the rich nations complain that there is no point in cutting their own emissions if emissions are to continue to grow in China and India. The governments of China and India complain that limiting their pollution is a waste of time if the richer countries – whose output per head is still far greater than theirs – are not prepared to make the necessary reductions. It is also true that the fossil fuel companies use their tremendous wealth to buy everything they need, including a politician's suit with the politician still inside it.

But if those governments that have expressed a commitment to stopping climate change have found their efforts frustrated, it is partly because they wanted them to be frustrated. They know that inside their electors there is a small but insistent voice asking them both to try and to fail. They know that if they had the misfortune to succeed, our lives would have to change. They know that we can contemplate a transformation of anyone's existence but our own.

So they play to the script which we have all ghost-written. They will make frowning speeches about the threat to the planet and the need for action. They will announce that this issue is of such importance that it transcends the usual political differences and requires a cross-party consensus. They will urge everyone to pull together and confront the enormity of the threat. Then they will discover, to their great disappointment, that progress has not been made, that it is in fact very difficult to make, and the decision about what should be done will yet again have to be deferred.

In the United Kingdom, as my researcher Matthew Prescott pointed out to me, government policy is not contained within the reports and reviews it commissions; government policy *is* the reports and reviews. By commissioning endless inquiries into the problem and the means by which it might be tackled, the government creates the impression that something is being done, while simultaneously preventing anything from happening until the next review (required to respond to the findings of the last review) has been published. I have an image in my mind of the British prime minister up to his neck in water on the floor of the House of Commons, explaining that 'in the forth-coming White Paper on energy efficiency . . .'.

Governments will pursue this course of inaction – irrespective of the human impacts – while it remains politically less costly than the alternative. The task of climate-change campaigners is to make it as expensive as possible. This means abandoning the habit of mind into which almost all of us have somehow slumped over the past ten years or so: the belief that someone else will do it for us.

I am not entirely sure how this came about. In the early 1990s, activists lay in the road and sat in trees and on the roofs of ministers' houses and interrupted their speeches and poured fake blood on the steps of Downing Street and made such a thoroughgoing nuisance of themselves that, however hard politicians tried to block them out, eventually they had to be heard. Now we sit on our broadening backsides and moan about the fact that everyone else is moaning about the fact and not *doing* anything.

This is partly, I think, because of the sustained global economic growth between then and now. We are simply too comfortable, and we have too much to lose. It is partly also because, accompanying this growth (indeed to some extent driving it) has been a surge in indebtedness, especially among the young, who used to be on the front line. Debt induces a bright panic, which ensures that those burdened with it can seldom see beyond the next few weeks.

But I also blame that tool of empowerment, the internet. Of course it is marvellously useful, allows us to exchange information, find the facts we need, alert each other to the coming dangers and all the rest of it. But it also creates a false impression of action. It allows us to believe that we can change the world without leaving our chairs. We are being heard! Our voices resonate around the world, provoking commentary and debate, inspiring some, enraging others. Something is happening! A movement is building! But by itself, as I know to my cost, writing, reading, debate and dissent change nothing. They are of value only if they inspire action. Action means moving your legs. Indeed, if this book has not encouraged you to want to *do* something, then I urge you to return it to the shop and demand your money back, for it has proved to be useless.

But do what? On the next pages is a list of the names and addresses of the organisations already campaigning against climate change and the activities which cause it. I want you to join them. I want you to set up your own group only if they turn out to be going nowhere: I have seen plenty of movements disintegrate through proliferation. I want you to find out how you can be most useful to them. But above all, I want you to make an imaginative leap seldom demanded of you by governments or advertisers or newspapers or teachers.

For the campaign against climate change is an odd one. Unlike almost all the public protests which have preceded it, it is a campaign not for abundance but for austerity. It is a campaign not for more freedom but for less. Strangest of all, it is a campaign not just against other people, but also against ourselves.

Organizations Campaigning to Reduce the Impact of Climate Change

Stop Climate Chaos
The Grayston Centre
28 Charles Square
London N1 6HT
020 7324 4750
info@stopclimatechaos.org
http://www.stopclimatechaos.org

Climate Outreach Information Network
16B Cherwell Street
Oxford
OX4 1BG
01865 727911
info@coinet.org.uk
http://www.coinet.org.uk

Campaign against Climate Change
Top Floor
5 Caledonian Road
London N1 9DX
020 7833 9311
0790 3316331
campaigncc.org
http://www.campaigncc.org

Friends of the Earth
26–28 Underwood Street
London N1 7JQ
020 7490 1555
e-mail via http://www.foe.co.uk/feedbackcomment.html
http://www.foe.co.uk

Greenpeace
Canonbury Villas
London N1 2PN
020 7865 8100
info@uk.greenpeace.org
http://www.greenpeace.org.uk

World Wildlife Federation
Panda House
Weyside Park
Godalming
Surrey GU7 1XR
01483 426333
http://www.wwf.org.uk

New Economics Foundation
3 Jonathan Street
London SE11 5NH
020 7820 6300
info@neweconomics.org
http://www.neweconomics.org

People and Planet
51 Union Street
Oxford OX4 1JP
01865 245678
e-mail via http://peopleandplanet.org/aboutus/contact.php
http://peopleandplanet.org

World Development Movement
25 Beehive Place
London
SW9 7QR
Tel: 020 7737 6215
e-mail via http://www.wdm.org.uk/about/contact/index.htm
http://www.wdm.org.uk

Transport 2000
The Impact Centre
12–18 Hoxton Street
London N1 6NG
020 7613 0743
info@transport2000.org.uk
http://www.transport2000.org.uk

Road Block
PO Box 164
Totnes TQ9 5WX
020 7729 6973
07854 693067
office@roadblock.org.uk (phone preferred)
http://www.roadblock.org.uk

AirportWatch
Broken Wharf House
2 Broken Wharf
London EC4V 3DT
020 7248 2227
info@airportwatch.org.uk
http://www.airportwatch.org.uk

Centre for Alternative Technology
Machynlleth
Powys SY20 9AZ
01654 705950
http://www.cat.org.uk

Association for the Conservation of Energy
Westgate House
2a Prebend Street
London N1 8PT
020 7359 8000
info@ukace.or
http://www.ukace.org

Notes

INTRODUCTION:
THE FAILURE OF GOOD INTENTIONS

1. Christopher Marlowe, *Doctor Faustus* (W. W. Norton & Co, New York, 2005 [1604]), p. 20.
2. George Monbiot, 'Climate Change: A Crisis of Collective Denial?', the Professor David Hall Lecture, given to the Environmental Law Foundation at the Law Society, 4 May 2005. A transcript is available at
http://www.elflaw.org/files/David%20Hall%20Lecture%202005%20transcript.doc
3. Ian McEwan, *Saturday* (Jonathan Cape, London, 2005), pp. 149–50.
4. Colin Forrest, 'The Cutting Edge: Climate Science to April 2005':
http://www.climate-crisis.net/downloads/THE_CUTTING_EDGE_CLIMATE_SCIENCE_TO_APRIL_05.pdf
5. Energy Information Administration, *International Energy Annual 2003*, 2005, Table H.1cco2 (World Per Capita Carbon Dioxide Emissions from the Consumption and Flaring of Fossil Fuels, 1980–2003):
http://www.eia.doe.gov/pub/international/iealf/tableh1cco2.xls
6. ibid.
7. Martha Buckley, 'Bling Bling Merrily on High', *BBC News Online*, 23 December 2004: http://news.bbc.co.uk/2/hi/uk_news/4116563.stm
8. Craig McLean, 'The Importance of Being Earnest', *Guardian*, 28 May 2005.
9. ibid.
10. Bob Flowerdew, *Organic Bible* (Kyle Cathie Ltd, London, 2003), p. 17.
11. ibid., p. 178.
12. Bill Dunster Architects, 'Zedupgrade: An Introduction to Refurbishment Systems for Existing Homes', 19 March 2005:
http://www.zedfactory.com/ZEDupgrade_A4_Brochure.pdf
13. The Windsave 1000 system: http://www.windsave.com/WS1000.htm
14. Derek Taylor, 'Potential Outputs from 1–2m Diameter Wind Turbines', *Building for a Future*, winter 2005/6, special wind-power feature. This is extracted from the graph, and describes output at an average annual windspeed of 4 metres per second. The previous article in the same edition, by Nick Martin, explains that, in built-up

areas, 'Very few installations are likely to experience more than the equivalent of 4 metres per second average windspeed.'

15. Nick Martin, 'Can We Harvest Useful Wind Energy from the Roofs of Our Buildings?' *Building for a Future*, winter 2005/6, special wind-power feature, Table 2.

16. Godfrey Boyle, Bob Everett and Janet Ramage (eds.), *Energy Systems and Sustainability* (Oxford University Press, Oxford, 2003), Table 3.1, p. 104.

17. ibid., p. 104.

1 A FAUSTIAN PACT

1. *Doctor Faustus*, p. 25.

2. Quoted by Stuart Atkins, 'Survey of the Faust Theme', in Cyrus Hamlin (ed.), *Faust: A Norton Critical Edition* (W. W. Norton & Co, New York, 2001), p. 573.

3. ibid.

4. ibid., p. 574.

5. *Doctor Faustus*, Prologue, p. 7.

6. ibid., Act 1, Scene 1, p. 9.

7. ibid., Act 1, Scene 1, p. 10.

8. ibid., Act 1, Scene 3, p. 17.

9. ibid., Act 2, Scene 1, p. 25.

10. anon., 1589? *The History of the Damnable Life and Deserved Death of Doctor John Faustus* (published by Thomas Orwin in 1592), Chapter 2. Quoted in *Doctor Faustus*, p. 185.

11. J. U. Nef, *The Rise of the British Coal Industry* (Routledge, London, 1932). Cited by Godfrey Boyle, Bob Everett and Janet Ramage (eds.), *Energy Systems and Sustainability* (Oxford University Press, Oxford, 2003), p. 160.

12. US Energy Information Administration, 'Country Analysis Brief: United Kingdom', April 2005: http://www.eia.doe.gov/emeu/cabs/uk.html

13. ibid.

14. Urs Siegenthaler *et al.*, 'Stable Carbon Cycle–Climate Relationship During the Late Pleistocene', *Science*, Vol. 310 (25 November 2005), pp. 1313–17.

15. Renato Spahni *et al.*, 'Atmospheric Methane and Nitrous Oxide of the Late Pleistocene from Antarctic Ice Cores', *Science*, Vol. 310 (25 November 2005), pp. 1317–21.

16. Siegenthaler *et al.*, op. cit.

17. The Intergovernmental Panel on Climate Change notes that 'the rate of increase over the past century is unprecedented, at least during the past 20,000 years.' Intergovernmental Panel on Climate Change, *Climate Change 2001: Working Group I – The Scientific Basis*, Observed Changes in Globally Well-Mixed Greenhouse Gas Concentrations and Radiative Forcing:
http://www.grida.no/climate/ipcc_tar/wg1/016.htm

18. Department for Environment, Food and Rural Affairs, 'Methane Emissions by Source: 1970–2003', 23 January 2006:

http://www.defra.gov.uk/environment/statistics/globatmos/kf/gakf08.htm

19. Intergovernmental Panel on Climate Change, *Summary for Policymakers to Climate Change 2001: Synthesis Report of the IPCC Third Assessment Report* (IPCC, London, 2001), p. 4.

20. World Meteorological Organization, 'Extreme Weather Events Might Increase' (press release), 2 July 2003: http://www.wmo.ch/web/Press/Press695.doc

21. Tim P. Barnet *et al.*, 'Penetration of Human-induced Warming into the World's Oceans', *Science*, Vol. 309 (8 July 2005), pp. 284–7.

22. Tim Barnet, quoted in Fred Pearce, 'Climate Evidence Finds Us Guilty as Charged', *New Scientist*, 11 June 2005.

23. Naomi Oreskes, 'The Scientific Consensus on Climate Change', *Science*, Vol. 306 (3 December 2004), p. 1686. The original essay gave the search term as 'climate change', but this was corrected to 'global climate change' in a subsequent edition.

24. ibid.

25. The Royal Society *et al.*, 'The Science of Climate Change', *Science*, Vol. 292 (18 May 2001), p. 1261.

26. National Academy of Science, *Climate Change Science: An Analysis of Some Key Questions* (National Academy Press, Washington, DC, 2001): http://fermat.nap.edu/html/climatechange/

27. American Meteorological Society, 'Climate Change Research: Issues for the Atmospheric and Related Sciences' (approved by AMS Council, 9 February 2003), *Bulletin of the American Meteorological Society*, April 2003, pp. 508–15.

28. American Geophysical Union Council, 'AGU Position Statement on Human Impacts on Climate', *Eos*, Vol. 84 (December 2003), p. 574.

29. American Association for the Advancement of Science, 'AAAS Atlas of Population and Environment: Climate Change', 2000: http://www.ourplanet.com/aaas/pages/atmos02.html

30. Roy W. Spencer and John R. Christy, 'Precision and Radiosonde Validation of Satellite Gridpoint Temperature Anomalies. Part II: a Tropospheric Retrieval and Trends During 1979–90', *Journal of Climate*, Vol. 5 (1992), pp. 858–66.

31. Carl A. Mears and Frank J. Wentz, 'The Effect of Diurnal Correction on Satellite-Derived Lower Tropospheric Temperature', *Science*, Vol. 309 (2 September 2005), pp. 1548–51.

32. B.D. Santer *et al.*, 'Amplification of Surface Temperature Trends and Variability in the Tropical Atmosphere', *Science*, Vol. 309 (2 Septermber 2005), pp. 1548–51.

33. Steven J. Sherwood, John R. Lanzante and Cathryn L. Meyer, 'Radiosonde Daytime Biases and Late Twentieth-Century Warming', *Science*, Vol. 309 (2 September 2005), pp. 1556–9.

34. Steven J. Sherwood, quoted by Zeeya Merali, 'Sceptics Forced into Climate Climb-down', *New Scientist*, 20 August 2005.

35. National Snow and Ice Data Centre, 'Sea Ice Decline Intensifies', press release, 28 September 2005: http://nsidc.org/news/press/20050928_trendscontinue.html

36. The collapse is described by the scientists who saw it in an article by John Vidal,

'Antarctica Sends 500 Million Billion Tonne Warning of the Effects of Global Warming', *Guardian*, 20 March 2006.

37. Andrew Shepherd, Duncan Wingham, Tony Payne, Pedro Skvarca, 'Larsen Ice Shelf Has Progressively Thinned', *Science*, Vol. 302 (31 October 2003), pp. 856–8.

38. Intergovernmental Panel on Climate Change, op. cit. (note 19, above), p. 12.

39. Meteorological Office, *International Symposium on the Stabilisation of Greenhouse Gases: Tables of Impacts* (Hadley Centre, Exeter, 2005), Table 3 (Major Impacts of Climate Change on the Earth System):
http://www.stabilisation2005.com/impacts/impacts_earth_system.pdf

40. World Glacier Monitoring Service, *Glacier Mass Balance Data 2004*, 2006:
http://www.geo.unizh.ch/wgms/mbb/mb04/sum04.html

41. Fred Pearce, 'Climate Warning as Siberia Melts', *New Scientist*, 11 August 2005.

42. Erica Goldman, 'Even in the High Arctic, Nothing is Permanent', *Science*, Vol. 297 (30 August 2002), pp. 1493–4.

43. Meteorological Office, op. cit. (note 39, above).

44. World Health Organization, *Climate Change*, 2003:
http://www.who.int/heli/risks/climate/climatechange/en/index.html

45. Intergovernmental Panel on Climate Change, op. cit. (note 19, above), p. 5.

46. Daniel A. Stainforth *et al.*, 'Uncertainty in Predictions of the Climate Response to Rising Levels of Greenhouse Gases', *Nature*, Vol. 433 (27 January 2005), pp. 403–6.

47. Martin Parry *et al.*, 'Millions at Risk: Defining Critical Climate Change Threats and Targets', *Global Environmental Change*, Vol. 11 (2001), pp. 181–3.

48. Meteorological Office, op. cit. (note 39, above), Table 2a (Impacts on Human Systems Due to Temperature Rise, Precipitation Change and Increases in Extreme Events): http://www.stabilisation2005.com/impacts/impacts_human.pdf

49. H. Schroder *et al.*, *Assessment of Renewable Ground and Surface Water Resources and the Impact of Economic Activity on Runoff in the Basin of the Ili River, Republic of Kazakhstan* (Kazakh Academy of Sciences, Almaty, Kazakhstan, 2002); and P. Wagnon, *et al.*, 1999, 'Energy Balance and Runoff Seasonality of a Bolivian Glacier', *Global and Planetary Change*, 22(1–4) (1999), pp. 49–58; both cited in World Wildlife Fund, 'Going, Going, Gone: Climate Change and Global Glacier Decline', 2003: http://assets.panda.org/downloads/glacierspaper.pdf

50. Intergovernmental Panel on Climate Change, *Climate Change 2001: Working Group II – Impacts, Adaptation and Vulnerability*:
http://www.grida.no/climate/ipcc_tar/wg2/005.htm

51. T. N. Palmer and J. Raisanen, 'Quantifying the Risk of Extreme Seasonal Precipitation Events in a Changing Climate', *Nature*, Vol. 415 (31 January 2002), pp. 512–14.

52. Günther Fischer, Mahendra Shah, Harrij van Velthuizen, and Freddy O. Nachtergaele, *Global Agro-ecological Assessment for Agriculture in the 21st Century* (International Institute for Applied Systems Analysis and the Food and Agriculture Organisation, July 2001):
http://www.iiasa.ac.at/Research/LUC/SAEZ/pdf/gaez2002.pdf

53. Julia M. Slingo, Andrew J. Challinor, Brian J. Hoskins and Timothy R. Wheeler,

'Introduction: Food Crops in a Changing Climate', *Philosophical Transactions of the Royal Society*, Vol. 360 (29 November 2005), pp. 1983–9.

54. Shaobing Peng *et al.*, 'Rice Yields Decline with Higher Night Temperature from Global Warming', *Proceedings of the National Academy of Sciences*, Vol. 101 (28 June 2004), pp. 9971–5.

55. Günther Fischer *et al.*, op. cit. (note 52, above).

56. Stephen P. Long, Elizabeth A. Ainsworth, Andrew D. B. Leakey and Patrick B. Morgan, 'Global Food Insecurity', *Philosophical Transactions of the Royal Society*, Vol. 360 (29 November 2005), pp. 2011–20.

57. ibid.

58. ibid.

59. Martin Parry, Cynthia Rosenzweig and Matthew Livermore, 'Climate Change, Global Food Supply and Risk of Hunger', *Philosophical Transactions of the Royal Society*, Vol. 360 (29 November 2005), pp. 2125–38.

60. Julia M. Slingo *et al.*, op. cit. (note 53, above).

61. Meteorological Office, op. cit. (note 48, above).

62. Jonathan A. Patz, Diarmid Campbell-Lendrum, Tracey Holloway and Jonathan A. Foley, 'Impact of Regional Climate Change on Human Health', *Nature*, Vol. 438 (17 November 2005), pp. 310–17.

63. Intergovernmental Panel on Climate Change, op. cit. (note 50, above).

64. Jonathan A. Patz *et al.*, op. cit. (note 62, above).

65. Conference of the International Association of Hydrogeologists, reported by Fred Pearce, 'Cities May Be Abandoned as Salt Water Invades', *New Scientist*, 16 April 2005.

66. ibid.

67. Intergovernmental Panel on Climate Change, op. cit. (note 50, above).

68. Professor Chris Rapley, director of the British Antarctic Survey, gave this warning in his presentation to the American Association for the Advancement of Science on 19 February 2006, *West Antarctic Ice Sheet: Waking the Sleeping Giant?*: http://www.eurekalert.org/pub_releases/2006–02/bas-wai021406.php

69. Intergovernmental Panel on Climate Change, *Climate Change 2001: Working Group I – The Scientific Basis*, Observed Changes in Climate Variability and Extreme Weather and Climate Events: http://www.grida.no/climate/ipcc_tar/wg1/014.htm

70. P. J. Webster, G. J. Holland, J. A. Curry and H. R. Chang, 'Changes in Tropical Cyclone Number, Duration and Intensity in a Warming Environment', *Science*, Vol. 309 (16 September 2005), pp. 1844–6.

71. Kerry Emanuel, 'Increasing Destructiveness of Tropical Cyclones Over the Past 30 Years', *Nature*, Vol. 436 (4 August 2005), pp. 686–8.

72. P. J. Webster *et al.*, op. cit. (note 70, above).

73. Cited by the House of Lords Select Committee on Economic Affairs on 6 July 2005, *The Economics of Climate Change*, Vol. I: Report (Stationery Office, London, 2005), p. 24.

74. Peter A. Stott, D. A. Stone and M. R. Allen, 'Human Contribution to the European Heatwave of 2003', *Nature*, Vol. 432 (2 December 2004), pp. 610–14.

75. ibid.

76. Intergovernmental Panel on Climate Change, op. cit. (note 50, above).

77. Chris D. Thomas *et al.*, 'Extinction Risk from Climate Change', *Nature*, Vol. 427 (8 January 2004), pp. 145–8.

78. Meteorological Office, op. cit. (note 39, above), Table 1a (Impacts of Level of Temperature Change on Ecosystems): http://www.stabilisation2005.com/impacts/impacts_ecosystems.pdf

79. ibid.

80. The Royal Society, 'Ocean Acidification Due to Increasing Atmospheric Carbon Dioxide', Policy Document 12/05 (June 2005): http://www.scar.org/articles/Ocean_Acidification(1).pdf

81. Meteorological Office, op. cit. (note 78, above).

82. Sharon A. Cowling *et al.*, 'Contrasting Simulated Past and Future Responses of the Amazonian Forest to Atmospheric Change', *Philosophical Transactions of the Royal Society*, Vol. 359 (29 March 2004), pp. 539–47.

83. ibid.

84. ibid.

85. P. M. Cox, C. Huntingford, and C. D. Jones, 'Conditions for Sink-to-Source Transitions and Runaway Feedbacks from the Land Carbon-Cycle', In H. J. Schellnhuber *et al.* (eds.), *Avoiding Dangerous Climate Change* (Cambridge University Press, Cambridge, 2006), pp. 155–61.

86. Chris D. Jones *et al.*, 'Strong Carbon Cycle Feedbacks in a Climate Model with Interactive CO_2 and Sulphate Aerosols', *Geophysical Research Letters*, Vol. 30 (9 May 2003), p. 1479.

87. Geoff Jenkins, 'A Question of Probability: Latest Research Identifies New Factors in Climate Change', *New Economy* (Institute for Public Policy Research, London), Vol. 10 (March 2003), pp. 144–9.

88. A. Angert *et al.*, 'Drier Summers Cancel Out the CO_2 Uptake Enhancement Induced by Warmer Springs', *Proceedings of the National Academy of Sciences*, Vol. 102 (2 August 2005), pp. 10823–7.

89. Pat H. Bellamy *et al.*, 'Carbon Losses from All Soils Across England and Wales 1978–2003', *Nature*, Vol. 437 (8 September 2005), pp. 245–8.

90. ibid.

91. Geoff Jenkins, op. cit. (note 87, above).

92. Fred Pearce, 'Climate Warning as Siberia Melts', *New Scientist*, 11 August 2005.

93. House of Lords Select Committee on Economic Affairs, op. cit. (note 73, above), p. 11.

94. For example, Environment Canada, 'The Science of Climate Change': http://www.msc.ec.gc.ca/education/scienceofclimatechange/understanding/FAQ/sections/2_e.html

95. National Center for Atmospheric Research, 'Most of Arctic's Near-Surface Permafrost May Thaw by 2100' (19 December 2005): http://www.ucar.edu/news/releases/2005/permafrost.shtml

96. For example, Andrew J. Weaver and Claude Hillaire-Marcel, 'Global Warming and the Next Ice Age', *Science*, Vol. 304 (16 April 2004), pp. 400–402.

97. House of Lords Select Committee on Economic Affairs, op. cit. (note 73, above), p. 26.

98. Detlef Quadfasel, 'The Atlantic Heat Conveyor Slows', *Nature*, Vol. 438 (1 December 2005), pp. 565–6.

99. Harry L. Bryden, Hannah R. Longworth and Stuart A. Cunningham, 'Slowing of the Atlantic Meridional Overturning Circulation at 25° N', *Nature*, Vol. 438 (1 December 2005), pp 655–7.

100. Detlef Quadfasel, op. cit. (note 98, above).

101. ibid.

102. Intergovernmental Panel on Climate Change, op. cit. (note 19, above): http://www.grida.no/climate/ipcc_tar/vol4/english/038.htm

103. Cited by Fred Pearce, 'Heat Will Soar as Haze Fades', *New Scientist*, 7 June 2003.

104. Daniel A. Stainforth *et al.*, op. cit. (note 46, above).

105. ibid.

106. Michael J. Benton, *When Life Nearly Died: The Greatest Mass Extinction of All Time* (Thames and Hudson, London, 2003).

107. ibid.

108. ibid.

109. Jeffrey T. Kiehl and Christine A. Shields, 'Climate Simulation of the Latest Permian: Implications for Mass Extinction', *Geology*, Vol. 33 (September 2005), pp. 757–60.

110. Michael J. Benton, op. cit. (note 106, above).

111. Jeffrey T. Kiehl and Christine A. Shields, op. cit. (note 109, above).

112. ibid.

113. For example, James Lovelock, 'The Earth is About to Catch a Morbid Fever that May Last as Long as 100,000 years', *Independent*, 16 January 2006.

114. Bill Hare, 'Relationship Between Increases in Global Mean Temperature and Impacts on Ecosystems, Food Production, Water and Socio-Economic Systems', in Hans Joachim Schellnhuber (ed.), *Avoiding Dangerous Climate Change* (Cambridge University Press, 2006), pp. 191–9: http://www.defra.gov.uk/environment/climatechange/internat/pdf/avoid-dangercc.pdf

115. Hartmut Grassl *et al.*, *Climate Protection Strategies for the 21st Century: Kyoto and Beyond*, WBGU Special Report (WBGU (German Advisory Council on Global Change), Berlin, 2003), p. 11: http://www.wbgu.de/wbgu_sn2003_engl.pdf

116. Meteorological Office, op. cit. (note 48, above).

117. ibid.

118. Meteorological Office, op. cit. (note 78, above).

119. Meteorological Office, op. cit. (note 48, above).

120. Meteorological Office, op. cit. (note 78, above).

121. Meteorological Office, op. cit. (note 39, above).

122. Colin Forrest extracts this date from the figures given in the paper by Chris D. Jones *et al.* (note 86, above). Colin Forrest, 'The Cutting Edge: Climate Science to April 2005':

http://www.climate-crisis.net/downloads/THE_CUTTING_EDGE_CLIMATE_SCIENCE_TO_APRIL_05.pdf

123. Bill Hare and Malte Meinshausen, *How Much Warming Are We Committed To and How Much Can Be Avoided?*, PIK report 93 (Potsdam Institute for Climate Impact Research, Potsdam, 2004), Figure 7, page 24:
http://www.pik-potsdam.de/publications/pik_reports/reports/pr.93/pr93.pdf

124. Colin Forrest, op. cit. (note 122, above).

125. Energy Information Administration, *International Energy Annual 2003*, 2005, Table H.1cco2 (World Per Capita Carbon Dioxide Emissions from the Consumption and Flaring of Fossil Fuels, 1980–2003):
http://www.eia.doe.gov/pub/international/iealf/tableh1cco2.xls

126. The figures (taken from Energy Information Administration, op. cit. (note 125, above) work out as follows:

country	carbon dioxide emissions per capita, 2003 (tonnes)	carbon emissions per capita (CO_2 3.667) (tonnes)	percentage cut required
France	6.8	1.9	83
UK	9.5	2.6	87
Germany	10.2	2.8	88
Canada	19.1	5.2	94
Australia	19.1	5.2	94
US	20.0	5.5	94

127. Bill Hare and Malte Meinshausen, op. cit. (note 123, above), p. 24.

128. Paul Baer and Tom Athanasiou, 'Honesty About Dangerous Climate Change', 2005: http://www.ecoequity.org/ceo/ceo_8_2.htm

129. The Royal Society, 'A Guide to Facts and Fictions about Climate Change', 2005: http://www.royalsoc.ac.uk/downloaddoc.asp?id=1630

130. Johann Wolfgang von Goethe, *Faust* (1770–1830), published in translation by Cyrus Hamlin (ed.), *Faust*, Norton Critical Edition (W. W. Norton and Co., New York, 2001), Part 1, Study, line 1695.

131. ibid., Part 1, Study, lines 1765–8.

132. ibid., Part II, Act V, line 11563.

2 THE DENIAL INDUSTRY

1. *Faust*, lines 11693–6.

2. Dr Andrew Dlugolecki, formerly director of general insurance development at CGNU, at the 'Decarbonising the UK' conference, 21 September 2005, Church House, Westminster.

3. Intergovernmental Panel on Climate Change, *Climate Change 2001: Working Group II – Impacts, Adaptation and Vulnerability*, Table 11–9:
http://www.grida.no/climate/ipcc_tar/wg2/446.htm

4. Alan Dupont and Graeme Pearman, 'Heating up the Planet: Climate Change and Security' (Paper 12, The Lowy Institute, 13 June 2006), pp. 45–6:
http://www.lowyinstitute.org/Publication.asp?pid=391

5. anon., 'Washed Away', *New Scientist*, 25 June 2005.

6. James Verdin, Chris Funk, Gabriel Senay and Richard Choularton, 'Climate Science and Famine Early Warning', *Philosophical Transactions of the Royal Society*, Vol. 360 (29 November 2005), pp. 2155–68.

7. ibid.

8. See also Günther Fischer, Mahendra Shah, Harrij van Velthuizen and Freddy O. Nachtergaele, July 2001. 'Global Agro-ecological Assessment for Agriculture in the 21st Century', International Institute for Applied Systems Analysis and Food and Agriculture Organisation, July 2001:
http://www.iiasa.ac.at/Research/LUC/SAEZ/pdf/gaez2002.pdf

9. Energy Information Administration, *International Energy Annual 2003*, 2005, Table H.1cco2 (World Per Capita Carbon Dioxide Emissions from the Consumption and Flaring of Fossil Fuels, 1980–2003):
http://www.eia.doe.gov/pub/international/iealf/tableh1cco2.xls

10. *The Economist Pocket World in Figures 2005* (Profile Books Ltd, London, 2004), p. 28.

11. Tony Blair, speech on climate change, 14 September 2004:
http://www.number10.gov.uk/output/Page6333.asp

12. Peter Hitchens, 'Global Warming? It's Hot Air and Hypocrisy', *Mail on Sunday*, 29 July 2001.

13. Melanie Phillips, 'Does This Prove that Global Warming's All Hot Air?', *Daily Mail*, 13 January 2006.

14. Peter Hitchens, op. cit. (note 12, above).

15. Melanie Phillips, speaking on *The Moral Maze*, BBC Radio 4, 17 February 2005.

16. David Bellamy, 'Global Warming? What a Load of Poppycock!' *Daily Mail*, 9 July 2004.

17. David Bellamy, letter to *New Scientist*, 16 April 2005.

18. Conversation with Dr Frank Paul of the World Glacier Monitoring Service, 5 May 2005.

19. He cited Frank Paul *et al.*, 'Rapid Disintegration of Alpine Glaciers Observed with Satellite Data', *Geophysical Research Letters*, Vol. 31 (12 November 2004), L21402; and WGMS, 'Fluctuations of Glaciers 1990–1995 Vol. VII', 1988:
http://www.wgms.ch/fog/fog7.pdf. A fuller list of recent publications on glacial movements and mass balance is available at
http://www.wgms.ch/literature.html

20. E-mail from David Bellamy, 5 May 2005.

21. http://www.iceagenow.com/Growing_Glaciers.htm

22. Roger Boyes, 'Blame the Jews', *The Times*, 7 November 2003.

23. David Bamford, 'Turkish Officials Carpeted', *Guardian*, 30 July 1987.

24. Michael White, 'Will the Democrats Wear this Whig?' *Guardian*, 3 May 1986.

25. Francis Wheen, 'Branded: Lord Rees-Mogg, International Terrorist', *Guardian*, 21 August 1996.

26. Extract from Chip Berlet and Matthew N. Lyons, *Right-Wing Populism in America: Too Close for Comfort* (Guilford Press, New York, 2000), republished at http://www.publiceye.org/larouche/synthesis.html

27. This is the constant theme of *21st Century Science and Technology*.

28. Terry Kirby, 'The Cult and the Candidate', *Independent*, 21 July 2004.

29. Chip Bertlet, 20 December 1990:
http://www.skepticfiles.org/socialis/woo_left.htm

30. http://www.cei.org/gencon/014,02867.cfm, viewed 23 May 2006.

31. http://www.nationalcenter.org/NPA218.html, viewed 23 May 2006.

32. http://www.junkscience.com/nov98/moore.htm, viewed 23 May 2006.

33. John K. Carlisle (a director of the National Center for Public Policy Research), letter to the *Washington Post*, 17 November 1998.

34. http://www.sepp.org/controv/glaciers.html, viewed 7 May 2005.

35. George Monbiot, 'Junk Science', *Guardian*, 10 May 2005.

36. E-mail from Ron Partridge to S. Fred Singer, 11 May 2005.

37. E-mail from S. Fred Singer to Ron Partridge, 11 May 2005.

38. Second e-mail from Ron Partridge to S. Fred Singer, 11 May 2005.

39. Further e-mail from S. Fred Singer, forwarded to me by Ron Partridge, 28 May 2005.

40. http://www.sepp.org/pressrel/goreglac.html, viewed 23 May 2006.

41. David Teather, 'Washington Focuses on Oil Profits', *Guardian*, 7 November 2005.

42. Frank Luntz, 'The Environment: A Cleaner, Safer, Healthier America', 2002. The leaked memo can be seen here:
http://www.ewg.org/briefings/luntzmemo/pdf/LuntzResearch_environment.pdf

43. www.exxonsecrets.org

44. Quoted by Zeeya Merali, 'Sceptics Forced into Climate Climbdown', *New Scientist*, 20 August 2005.

45. The petition can be read here:
http://www.oism.org/pproject/s33p37.htm

46. Letter from Frederick Seitz, 'Research Review of Global Warming Evidence', 1998: http://www.oism.org/pproject/s33p41.htm

47. PRWatch, 'Case Study: The Oregon Petition", no date:
http://www.prwatch.org/improp/oism.html

48. http://www.exxonsecrets.org/html/orgfactsheet.php?id=36

49. PRWatch, op. cit. (note 47, above).

50. Arthur B. Robinson, Sallie L. Baliunas, Willie Soon and Zachary W. Robinson, 'Environmental Effects of Increased Atmospheric Carbon Dioxide', Oregon Institute of Science and Medicine and the George C. Marshall Institute, 1998:
http://www.oism.org/pproject/s33p36.htm

51. See the entries for Sallie Baliunas and Willie Soon at www.exxonsecrets.org

52. John H. Cushman Jr, 'Industrial Group Plans to Battle Climate Treaty', *New York Times*, 26 April 1998.

53. Arthur B. Robinson *et al.*, op. cit. (note 50, above).

54. National Academy of Sciences, 'Statement by the Council of the National Academy of Sciences Regarding Global Change Petition', 20 April 1998:
http://www4.nationalacademies.org/news.nsf/isbn/s04201998?OpenDocument

55. Environmental Protection Agency, December 1992, *Respiratory Health Effects Of Passive Smoking: Lung Cancer And Other Disorders*, EPA/600/6–90/006F (US Environmental Protection Agency, Washington, DC, December 1992), p. 21.

56. Ellen Merlo, memo to William I. Campbell, 17 February 1993, Bates no. 2021183916–2021183925, p. 1:
http://legacy.library.ucsf.edu/cgi/getdoc?tid=qdf02a00&fmt=pdf&ref=results

57. ibid., p. 5.

58. ibid., pp. 5–6.

59. ibid., p. 6.

60. ibid., p. 7.

61. ibid., p. 9.

62. Ted Lattanzio, note for Tina Walls, 20 May 1993, Bates no. 2021178204:
http://legacy.library.ucsf.edu/cgi/getdoc?tid=huj46e00&fmt=pdf&ref=results

63. Margery Kraus, letter to Vic Han, Director of Communications, Philip Morris USA, 23 September 1993, Bates no. 2024233698–2024233702, p. 2:
http://legacy.library.ucsf.edu/cgi/getdoc?tid=dqa35e00&fmt=pdf&ref=results

64. Tom Hockaday and Neal Cohen, memo to Matt Winokur, Director of Regulatory Affairs, Philip Morris, 'Thoughts on TASSC Europe', 25 March 1994, Bates no. 2024233595–2024233602, pp. 2–3:
http://legacy.library.ucsf.edu/cgi/getdoc?tid=pqa35e00&fmt=pdf&ref=results

65. Margery Kraus, op. cit. (note 63, above).

66. APCO Associates, 'Proposed Plan for the Public Launching of TASSC', 30 September 1993, Bates no. 2024233709–2024233717:
http://legacy.library.ucsf.edu/cgi/getdoc?tid=eqa35e00&fmt=pdf&ref=results

67. APCO Associates, 'Revised Plan for the Public Launching of TASSC', 15 October 1993, Bates no. 2045930493–2045930502:
http://legacy.library.ucsf.edu/cgi/getdoc?tid=aly03e00&fmt=pdf&ref=results

68. APCO Associates, op. cit. (note 66, above), p. 3

69. ibid.

70. For the authorship of these answers, see Jack Leonzi, note for Ellen Merlo, 15 November 1993, Bates no. 2024233664:
http://legacy.library.ucsf.edu/cgi/getdoc?tid=jqa35e00&fmt=pdf&ref=results

71. 'Draft Q and A for PM USA and TASSC', no date, Bates no. 2065556600:
http://legacy.library.ucsf.edu/cgi/getdoc?tid=ynk73c00&fmt=pdf&ref=results

72. ibid.

73. See David Michaels, 'Scientific Evidence and Public Policy', *American Journal of Public Health* (Supplement on Scientific Evidence and Public Policy), Vol. 95 (2005), pp. S5–S7.

74. anon., 'Smoking and Health Proposal' (Brown & Williamson), no date, Bates no. 690010951–690010959, p. 4:
http://legacy.library.ucsf.edu/cgi/getdoc?tid=rgy93f00&fmt=pdf&ref=results
75. http://www.exxonsecrets.org/html/orgfactsheet.php?id=6
76. Steven J. Milloy, 'Annual Report to TASSC Board Members', 7 January 1998, Bates no. 2065254885–2065254890, p. 3:
http://legacy.library.ucsf.edu/cgi/getdoc?tid=any77d00&fmt=pdf&ref=results
77. http://www.junkscience.com/nov98/moore.htm, viewed 23 May 2006.
78. http://www.junkscience.com/define.htm, viewed 24 May 2006.
79. http://www.sourcewatch.org/index.php?title=Steven_J._Milloy
80. ibid.
81. Tom Borelli, memo ('Junk Science'), Philip Morris Management Corporation, 11 April 1996, Bates no. 2505642662:
http://legacy.library.ucsf.edu/cgi/getdoc?tid=zna25c00&fmt=pdf&ref=results
82. Steven J. Milloy, op. cit. (note 76, above), p. 1.
83. Steven J. Milloy, op. cit. (note 76, above), p. 3.
84. Philip Morris, 'Public Policy Recommendations for 1997 with Paid Status', 25 February 1998, Bates no. 2063351196–2063351220, p. 4:
http://legacy.library.ucsf.edu/cgi/getdoc?tid=fah53a00&fmt=pdf&ref=results
85. Philip Morris, 'Issues Management', 2001, Bates no. 2082656417–2082656505, p. 13:
http://legacy.library.ucsf.edu/cgi/getdoc?tid=kwk84a00&fmt=pdf&ref=results
86. Paul D. Thacker, 'Pundit for Hire', *New Republic*, 26 January 2006.
87. Steven J. Milloy, 'Omitted Epidemiology', *British Medical Journal*, Vol. 317 (2 October 1998): http://bmj.bmjjournals.com/cgi/eletters/317/7163/903/a
88. 2004 IRS documents, cited by Environmental Science and Technology, 'The Junkman Climbs to the Top', 11 May 2005:
http://pubs.acs.org/subscribe/journals/esthag-w/2005/may/business/pt_junkscience.html
89. http://www.exxonsecrets.org/html/personfactsheet.php?id=881
90. ibid.
91. Paul D. Thacker, op. cit. (note 86, above).
92. Philip Morris, op. cit. (note 84, above), pp. 3–4.
93. Steven Milloy, 'Secondhand Smokescreen', 4 April 2001:
http://www.foxnews.com/story/0,2933,1897,00.html
94. Steven Milloy, 'Second-Hand Smokescreens', 4 June 2001:
http://www.foxnews.com/story/0,2933,26109,00.html
95. Steven Milloy, 'Kyoto's Quiet Anniversary', 16 February 2006:
http://www.foxnews.com/story/0,2933,185171,00.html
96. Steven Milloy, 'Hot Air Hysteria', 16 March 2006:
http://www.foxnews.com/story/0,2933,188176,00.html
97. Steven Milloy, 'The Greenhouse Myth', 20 April 2006:
http://www.foxnews.com/story/0,2933,192544,00.html
98. Paul D. Thacker, op. cit. (note 86, above).
99. For example

http://www.foxnews.com/story/0,2933,196118,00.html, viewed May 24 2006.

100. Steven J. Milloy, op. cit. (note 76, above).

101. Colin Stokes, Chairman of RJ Reynolds, 'RJR's Support of Biomedical Research', November 1979: Bates no. 504480506–504480517, p. 7: http://legacy.library.ucsf.edu/cgi/getdoc?tid=uyr65d00&fmt=pdf&ref=results

102. John L. Bacon, Director of Corporate Contributions, RJ Reynolds, interoffice memorandum ('Consultancy Agreements – Dr's Seitz and McCarty') to Edward A. Horrigan, Jr, Chairman and Chief Executive Officer, RJ Reynolds, 15 July 1986, Bates no. 508455416: http://tobaccodocuments.org/rjr/508455415–5416.html?pattern= 508455416images

103. Edward A. Horrigan, Jr, letter to Frederick Seitz, 15 July 1986, Bates no. 508263286: http://tobaccodocuments.org/rjr/508263286–3286.html

104. RJ Reynolds, 'Procedures for Managing and Progress Monitoring of RJ Reynolds Industries Support of Biomedical Research', no date, Bates no. 502130487: http://legacy.library.ucsf.edu/cgi/getdoc?tid=cva29d00&fmt=pdf&ref=results

105. John L. Bacon, minutes of the RJ Reynolds Medical Research meeting, 13 September 1979, Bates no. 504480459–504480464: http://tobaccodocuments.org/rjr/504480459–0464.pdf

106. Colin Stokes, op. cit. (note 101, above).

107. Philip Morris, untitled notes for a presentation, 1992, Bates no. 2024102283– 2024102287, p. 5: http://legacy.library.ucsf.edu/cgi/getdoc?tid=pfa35e00&fmt=pdf&ref=results

108. This memo, and several other details of Seitz's involvement with the tobacco industry was brought to light by Norbert Hirschhorn and Stella Aguinaga Bialous, 'Second-hand Smoke and Risk Assessment: What Was In It for the Tobacco Industry?', *Tobacco Control*, Vol. 10 (2001), pp. 375–82: http://tc.bmjjournals.com/cgi/content/full/10/4/375

109. Tom Hockaday, memo ('Opinion Editorials on Indoor Air Quality and Junk Science') to Ellen Merlo *et al.*, 8 March 1993, Bates no. 2021178205: http://legacy.library.ucsf.edu/cgi/getdoc?tid=iuj46e00&fmt=pdf&ref=results

110. S. Fred Singer, 'Junk Science at the EPA', 1993, Bates no. 2021178206– 2021178208, p. 2: http://legacy.library.ucsf.edu/cgi/getdoc?tid=cuj46e00&fmt=pdf&ref=results

111. Tom Hockaday and Neal Cohen, op. cit. (note 64, above), p. 5.

112. Philip Morris, op. cit. (note 84, above).

113. ibid.

114. http://news.bbc.co.uk/nol/shared/bsp/hi/live_events/forums/04/1091183999/ html/f_info.stm

115. *Today Programme*, 19 May 2005, BBC Radio 4: http://www.bbc.co.uk/radio4/today/l...al_20050519.ram

116. http://www.junkscience.com/Junkman.html

117. Senator James M. Inhofe (R-Okla), Senate Floor Statement: The Science of Climate Change, 28 July 2003: http://inhofe.senate.gov/pressreleases/climate.htm

118. ibid.

119. http://www.exxonsecrets.org/html/orgfactsheet.php?id=2

120. Philip Morris, op. cit. (note 84, above), p. 3.

121. Letter from Myron Ebell to Phil Cooney, published in 'White House Effect', *Harper's* magazine, May 2004.

122. Andrew C. Revkin, 'Bush Aide Softened Greenhouse Gas Links to Global Warming', *New York Times*, 8 June 2005.

123. ibid.

124. Jamie Wilson, 'Bush's Climate Row Aide Joins Oil Giant', *Guardian*, 16 June 2005.

125. A. G. (Randy) Randol III, Senior Environmental Adviser, ExxonMobil, memo ('Bush Team for IPCC negotiations') to John Howard, 6 February 2001, facsimile, sent from tel. no. (202) 8620268.

126. ibid., p. 2.

127. ibid., p. 5.

128. *Oxford English Dictionary*.

129. Sir David King, The Greenpeace Business lecture, 13 October 2004: http://www.greenpeace.org.uk/contentlookup.cfm?CFID=1322590&CFTOKEN=77800261&ucidparam=20041013100519

130. David King, speech to the 'Decarbonising the UK' conference, 21 September 2005, Church House, Westminster.

131. Simon Retallack, 'Setting a Long Term Climate Objective: A Paper for the International Taskforce on Climate Change' (Institute for Public Policy Research, October 2004): http://www.ippr.org.uk/ecomm/files/climate_objective.pdf

3 A RATION OF FREEDOM

1. *Doctor Faustus*, p. 30.

2. Richard Layard, *Happiness: Lessons from a New Science* (Allen Lane, London, 2005), p. 44.

3. George Orwell, 1940. 'The Lion and the Unicorn' (1940), in *Essays* (Penguin, London, 2000), p. 170.

4. Aubrey Meyer, *Contraction and Convergence: The Global Solution to Climate Change'*, Schumacher Briefing no. 5 (Green Books, Totnes, Devon, 2005).

5. See David Fleming, 'Energy and the Common Purpose: Descending the Energy Staircase with Tradeable Energy Quotas (TEQs)', no date: http://www.teqs.net/book/teqs.pdf

6. Tina Fawcett, presentation ('Personal Carbon Allowances and Industrial and Commercial Sector Capping') to the UK Energy Research Centre's workshop 'Taxing and Trading', 3 November 2005.

7. Richard Starkey and Kevin Anderson, *Domestic Tradable Quotas: A Policy Instrument for Reducing Greenhouse Gas Emissions from Energy Use*, Tyndall Centre Technical Report No. 39, December 2005.

8. See Kevin Smith *et al.*, *Hoodwinked in the Hothouse: The G8, Climate Change*

and Free Market Environmentalism, Transnational Institute Briefing Series, 30 June 2005: http://www.tni.org/reports/ctw/hothouse.pdf

9. Roger Harrabin, '£1bn Windfall from Carbon Trading', *BBC News Online*, 1 May 2006: http://news.bbc.co.uk/1/hi/sci/tech/4961320.stm

10. P. Ekins and S. Dresner, *Green Taxes and Charges: Reducing their Impact on Low-income Households* (Joseph Rowntree Foundation, York, 2004), cited in *Decarbonising the UK – Energy for a Climate Conscious Future* (The Tyndall Centre for Climate Change Research, 2005), p. 60:
http://www.tyndall.ac.uk/media/news/tyndall_decarbonising_the_uk.pdf

11. Roger Levett, 'Carbon Rationing Versus Energy Taxes: A False Opposition?', 25 October 2005:
http://www.ukerc.ac.uk/component/option,com_docman/task,doc_download/gid,325/

12. Bjørn Lomborg, *The Skeptical Environmentalist* (Cambridge University Press, Cambridge, 2001), p. 310.

13. ibid., p. 312.

14. Richard D. Knabb, Jamie R. Rhome, and Daniel P. Brown, 'Tropical Cyclone Report: Hurricane Katrina', National Hurricane Center, 20 December 2005:
http://www.nhc.noaa.gov/pdf/TCR-AL122005_Katrina.pdf

15. David Pearce *et al.*, 'The Social Costs of Climate Change: Greenhouse Damage and the Benefits of Control', in Intergovernmental Panel on Climate Change, *Climate Change 1995: Economic and Social Dimesnions of Climate Change* (Cambridge University Press, Cambridge, 1996), pp. 183–224.

16. Her Majesty's Treasury/Department for Environment, Food and Rural Affairs, 'Estimating the Social Cost of Carbon Emissions', 2002:
http://www.hm-treasury.gov.uk/media/209/60/scc.pdf

17. Performance and Innovation Unit, *The Energy Review*, February 2002, Annex 6, Table 1: http://www.number-10.gov.uk/su/energy/20.html

18. Commission of the European Communities, 'Winning the Battle against Global Climate Change', background paper, 9 February 2005, p. 25:
http://europa.eu.int/comm/press_room/presspacks/climate/staff_work_paper_sec_2005_180_3.pdf

19. Christian Azar and Stephen Schneider, 2002. 'Are the Economic Costs of Stabilizing the Atmosphere Prohibitive?', *Ecological Economics*, Vol. 42, pp. 73–80. See page 76.

20. ibid., p. 77.

21. Centrica, quoted by *BBC News Online*, 'Gas Prices "Set to Rise Further" ':
http://news.bbc.co.uk/1/hi/uk/4684108.stm

22. Christopher Adams, 'Industry Feels Heat of Gas Price Surge', *Financial Times*, 17 February 2006.

23. See, for example, 'Copenhagen Consensus, 2005. The Results':
http://www.copenhagenconsensus.com/Files/Filer/CC/Press/UK/copenhagen_consensus_result_FINAL.pdf

24. Organization for Economic Co-operation and Development, 'Aid from DAC Members, Donor Aid Charts', United States (for 2004), 2006:

http://www.oecd.org/dataoecd/42/30/1860571.gif

25. Organisation for Economic Co-operation and Development, 'Aid from DAC Members, Donor Aid Charts', United Kingdom (for 2004, 2006): http://www.oecd.org/dataoecd/42/53/1860562.gif

26. Department for Transport statistics, December 2005, collated by Road Block: http://www.roadblock.org.uk/press_releases/info/TPI%20and%20local% 20schemes%20Dec05.xls

27. Lord McKenzie of Luton, parliamentary answer HL 1508, 10 October 2005: http://www.publications.parliament.uk/pa/ld200405/ldhansrd/pdvn/lds05/text/ 51010w04.htm

28. Norman Myers and Jennifer Kent, *Perverse Subsidies: How Tax Dollars can Undercut the Environment and the Economy* (Island Press, Washington DC, 2001), p. 14.

29. ibid., p. 13.

30. http://energycommerce.house.gov/108/energy_pdfs_2.htm

31. Rep. Henry Waxman, letter to the Honorable J. Dennis Hastert, 27 July 2005: http://www.democrats.reform.house.gov/Documents/20050727165629-26334.pdf

32. ibid.

33. European Environment Agency, *Energy Subsidies in the European Union: A Brief Overview* (EEA, Copenhagen, 2004), p. 14: http://reports.eea.eu.int/technical_report_2004_1/en/Energy_FINAL_web.pdf

34. ibid., p. 9.

35. ibid., p. 14.

36. Linda Bilmes and Joseph E. Stiglitz, 'The Economic Costs of the Iraq War: An Appraisal Three Years after the Beginning of the Conflict', Working Paper Number RWP06–002, John F. Kennedy School of Government, Harvard University, 11 January 2006: http://ksgnotes1.harvard.edu/research/wpaper.nsf/rwp/RWP06–002/$File/ rwp_06_002_Bilmes_SSRN.pdf

37. For example, Energy Technology Support Unit, *New and Renewable Energy: Prospects in the UK for the 21st Century – Supporting Analysis* (ETSU, Harwell, 1999), pp. 227–8.

38. For example, http://www.peakoil.net, http://www.oilcrisis.com, and Matthew Simmons, *Twilight in the Desert: The Coming Saudi Oil Shock and the World Economy* (Wiley, New York, 2005).

39. For example, John H. Wood, Gary R. Long and David F. Morehouse, 'Long-Term World Oil Supply Scenarios: The Future Is Neither as Bleak or Rosy as Some Assert', Energy Information Administration, 18 August 2004: http://www.eia.doe.gov/pub/oil_gas/petroleum/feature_articles/2004/worldoil supply/oilsupply04.html

40. Robert L. Hirsch, Roger Bezdek and Robert Wendling, 'Peaking Of World Oil Production: Impacts, Mitigation, and Risk Management', US Department of Energy, February 2005, available at http://www.hubbertpeak.com/us/NETL/OilPeaking.pdf

41. ibid., p. 4.

42. ibid., p. 59.
43. International Energy Agency, 'Analysis of the Impact of High Oil Prices on the Global Economy', May 2004, p. 2:
http://www.iea.org/textbase/papers/2004/high_oil_prices.pdf
44. ibid.
45. *Doctor Faustus*, Act V, Scene 2.

4 OUR LEAKY HOMES

1. *Faust*, Lines 11604–5.
2. J. Daniel Khazzoom, 'Economic Implications of Mandated Efficiency Standards for Household Appliances', *Energy Journal*, Vol. 1 (1980), pp. 21–39.
3. Stanley Jevons, *The Coal Question – Can Britain Survive?* (1865), quoted in Horace Herring, 'Does Energy Efficiency Save Energy: The Implications of Accepting the Khazzoom–Brookes Postulate', April 1998:
http://technology.open.ac.uk/eeru/staff/horace/kbpotl.htm
4. J. Ausubel and H. D. Langford (eds.), *Technological Trajectories and the Human Environment* (US National Academy of Engineering, 1997), cited in *Decarbonising the UK – Energy for a Climate Conscious Future* (The Tyndall Centre for Climate Change Research, 2005), p. 70.
5. Paul Hawken, Amory B. Lovins and L. Hunter Lovins, *Natural Capitalism: The Next Industrial Revolution* (Earthscan, London, 1999), p. 62.
6. Katharina Kröger, Malcolm Fergusson and Ian Skinner, 'Critical Issues in Decarbonising Transport: The Role of Technologies', The Tyndall Centre for Climate Change Research, Working Paper 36, October 2003, p. 5:
http://tyndall.e-collaboration.co.uk/publications/working_papers/wp36.pdf
7. Roger Levett, 'Quality of Life Eco-Efficiency', *Energy and Environment*, Vol. 15 (2004), pp. 1015–26.
8. See Horace Herring, 'Does Energy Efficiency Save Energy: The Implications of Accepting the Khazzoom–Brookes Postulate', April 1998:
http://technology.open.ac.uk/eeru/staff/horace/kbpotl.htm
9. ibid.
10. Paul Hawken, Amory B. Lovins and L. Hunter Lovins, op. cit. (note 5, above), p. 13.
11. ibid., p. 126.
12. ibid.
13. Department of Foreign Affairs and Trade (government of Australia), 'Asia-Pacific Partnership on Clean Development and Climate', January 2006:
http://www.dfat.gov.au/environment/climate/ap6/
14. The latest regulations for existing dwellings are published by the Office of the Deputy Prime Minister, The Building Regulations 2000, Part L1B, 6 April 2006:
http://www.odpm.gov.uk/pub/338/ApprovedDocument
L1BConservationoffuelandpowerExistingdwellings2006edition_id1164338.pdf

15. House of Lords Select Committee on Science and Technology, 'Energy Efficiency', 5 July, 2005, paragraph 7.15:

http://www.publications.parliament.uk/pa/ld200506/ldselect/ldsctech/21/2102.htm

16. An official from the Department of Trade and Industry, speaking at the 'Resource '05' conference, Building Research Establishment, Watford, 15 September 2005.

17. Yvette Cooper, quoted by Paul Brown, 'Energy-saving Targets Scrapped', *Guardian*, 18 July 2005.

18. Department of Trade and Industry, 'Energy Consumption in the United Kingdom', 2003, p. 23: http://www.dti.gov.uk/files/file11250.pdf

19. Department of Trade and Industry, 'UK Energy Sector Indicators 2004', 2004, p. 97:

http://www.dti.gov.uk/energy/inform/energy_indicators/ind11_2004.pdf

20. Department of Trade and Industry, op. cit. (note 18, above), p. 11.

21. Department of Trade and Industry, op. cit. (note 19, above), p. 105.

22. Department of Trade and Industry, 'Energy: Its Impact on the Environment and Society', 2005, Chapter 3, page 9: http://www.dti.gov.uk/files/file20263.pdf

23. Brenda Boardman *et al.*, *40% House* (Environmental Change Institute, University of Oxford, 2005), p. 39.

24. House of Lords Select Committee on Science and Technology, op. cit. (note 15, above), paragraph 7.14.

25. Department of Trade and Industry, op. cit. (note 22, above), p. 9.

26. Paul Brown, op. cit. (note 17, above).

27. Andrew Warren, quoted in anon., 'Changes to Building Regulations "Substantially, Deliberately Weakened" ', *Energy World*, November/December 2005.

28. ibid.

29. David Olivier, 'Setting New Standards', *Building for a Future*, summer 2001.

30. House of Lords Select Committee on Science and Technology, op. cit. (note 15, above), Appendix 8.

31. David Strong, Building Research Establishment, evidence before the House of Lords Select Committee on Science and Technology, 9 February 2005, Question 534:

http://www.parliament.the-stationery-office.co.uk/pa/ld200506/ldselect/ldsctech/21/5020903.htm

32. Andrew Warren, 'Time to Put a Stop to the Disdain for Regulations', *Energy in Buildings and Industry*, March 2006:

http://www.ukace.org/pubs/articles/eibi2006-03.pdf

33. P. Grigg, 'Assessment of Energy Efficiency Impact of Building Regulations Compliance', report by the Building Research Establishment for the Energy Savings Trust and Energy Efficiency Partnership for Homes, 10 November 2004:

http://www.est.org.uk/uploads/documents/partnership/Houses_airtightness_report_Oct_04.pdf

34. David Strong, presentation to the 'Resource '05' conference, Building Research Establishment, Watford, 15 September 2005.

35. David Strong, op. cit. (note 31, above).

36. ibid.

37. House of Lords Select Committee on Science and Technology, op. cit. (note 15, above), para 6.25.

38. House of Commons Environmental Audit Committee, 'First Report', 19 January 2005, para 117:
http://www.publications.parliament.uk/pa/cm200405/cmselect/cmenvaud/135/13502.htm

39. Andrew Warren, Association for the Conservation of Energy, personal communication.

40. Mick Hamer, 'How Green is Your House?', *New Scientist*, 5 November 2005.

41. David Strong, op. cit. (note 31, above), Question 528.

42. Passiv Haus Institut, 'What is a Passive House?', no date: http://www.passiv.de

43. Jürgen Schnieders, 'CEPHEUS – Measurement Results from More Than 100 Dwelling Units in Passive Houses', May 2003:
http://www.passiv.de/07_eng/news/CEPHEUS_ECEEE.pdf

44. ibid.

45. ibid.

46. Patrick Bellew, Atelier 10, 13 October 2005, personal communication.

47. Peter Cox, 'Passivhaus', *Building for a Future*, winter 2005/6.

48. Passiv Haus Institut, op. cit. (note 42, above).

49. Jürgen Schnieders, op. cit. (note 43, above).

50. Peter Cox, op. cit. (note 47, above).

51. David Olivier, op. cit. (note 29, above).

52. H. F. Kaan and B. J. de Boer, 'Passive Houses: Achievable Concepts for Low CO_2 Housing', paper presented to the ISES conference 2005, Orlando, USA:
http://www.ecn.nl/docs/library/report/2006/rx06019.pdf

53. Jennie Organ, 'Beddington Zero Energy Development (BedZED)', Sustainable Development Commission, no date:
http://www.sd-commission.org.uk/communitiessummit/show_case_study.php/00035.html

54. Bedzed, no date. BedZED, no date, sales brochure:
http://www.bedzed.org.uk/BedZed_Brochure.pdf

55. The Royal Commission on Environmental Pollution, *Energy – The Changing Climate*, June 2000, Chapter 6, Box 6c:
http://www.rcep.org.uk/newenergy.htm

56. House of Commons Environmental Audit Committee, op. cit. (note 38, above).

57. House of Lords Select Committee on Science and Technology, op. cit. (note 15, above), para 7.28.

58. Brenda Boardman *et al.*, op. cit. (note 23, above), p. 88.

59. ibid., p. 39.

60. ibid., p. 43.

61. Hiroshi Matsumoto, 'System Dynamics Model for Life Cycle Assessment (LCA) of Residential Buildings', 1999:

http://www.ibpsa.org/%5Cproceedings%5CBS1999%5CBS99_PB-07.pdf

62. XCO2 Conisbee Ltd, 'Insulation for Sustainability – A Guide':
http://www.insulation.kingspan.com/newdiv/pdf/IfS%20Summary.pdf

63. House of Lords Select Committee on Science and Technology, op. cit. (note 15, above), para 7.2.

64. ibid.

65. Eoin Lees, 'Using Stamp Duty to bring about a Step Change in Household Energy Efficiency', report to CIGA/NIA and Association for the Conservation of Energy, 8 January 2005:
http://www.ukace.org/pubs/reportfo/2005%20Using%20Stamp%20Duty%20to%20bring%20about%20a%20Step%20Change%20in%20Household%20Energy%20Efficiency.pdf

66. Department for Environment, Food and Rural Affairs, 'Sustainable Energy: Energy Efficiency Commitment', 18 April 2006:
http://www.defra.gov.uk/environment/energy/eec/

67. Department of Trade and Industry, op. cit. (note 22, above), p. 11.

68. House of Lords Select Committee on Science and Technology, op. cit. (note 15, above), para 3.10.

69. HM Treasury, *Budget 2006*, Chapter 7, para 7.51:
http://www.hm-treasury.gov.uk/media/20F/1D/bud06_ch7_161.pdf

70. Brenda Boardman *et al.*, op. cit. (note 23, above), p. 48.

71. Department of Trade and Industry, op. cit. (note 22, above), p. 11.

72. Brenda Boardman *et al.*, op. cit. (note 23, above), p. 48.

73. Elliot Morley, parliamentary answer no. 4451, 16 June 2005:
http://www.publications.parliament.uk/pa/cm200506/cmhansrd/cm050616/text/50616w12.htmcolumn_564

74. House of Lords Select Committee on Science and Technology, op. cit. (note 15, above), para 9.26.

75. Brenda Boardman, 'Achieving Energy Efficiency through Product Policy: The UK Experience', *Environmental Science and Policy*, Vol. 7 (2004), pp. 165–76.

76. Department of Trade and Industry, op. cit. (note 22, above), p. 7.

77. Brenda Boardman *et al.*, op. cit. (note 23, above), p. 49.

78. ibid., p. 56.

79. Brenda Boardman, op. cit. (note 75, above).

80. House of Lords Select Committee on Science and Technology, op. cit. (note 15, above), para 9.4.

81. Friends of the Earth, 'Energy Saving Labels May Be Banned', press release, 20 October 2005:
http://www.foe.co.uk/resource/press_releases/energy_saving_labels_may_b_20102005.html

82. P. Schiellerup, 'An Examination of the Effectiveness of the EU Minimum Standard on Cold Appliances: The British Case', *Energy Policy*, Vol. 30 (2002), pp. 327–32.

83. House of Lords Select Committee on Science and Technology, op. cit. (note 15, above) para 9.7.

84. G. Wood and M. Newborough, 'Dynamic Energy-consumption Indicators for Domestic Appliances: Environment, Behaviour and Design', *Energy and Buildings*, Vol. 35 (2003), pp. 821–41.

85. L. McClelland and S. Cook, 'Energy Conservation Effects of Continuous In-home Feedback in All-electric Homes', *Journal of Environmental Systems*, Vol. 9 (1980), pp. 169–73; J. K. Dobson and J. D. Griffin, 'Conservation Effect of Immediate Electricity Cost Feedback on Residential Consumption Behaviour, in *Proceedings of the 7th ACEEE Summer Study on Energy Efficiency in Buildings* (ACEEE, Washington, DC, 1992); both cited in G. Wood and M. Newborough, op. cit. (note 84, above).

86. A. Meyel, *Low Income Households and Energy Conservation: Institutional, Behavioural and Housing Barriers to the Adoption of Energy Conservation Measures* (Built Environment Research Group, Polytechnic of Central London, London, 1987), cited in G. Wood and M. Newborough, op. cit. (note 84, above).

87. Ontario Energy Board, 'Smart Meter Initiative (RP-2004–0196)', 2005: http://www.oeb.gov.on.ca/html/en/industryrelations/ongoingprojects_ smartmeters.htm

88. House of Lords Select Committee on Science and Technology, op. cit. (note 15, above), para 5.57.

89. http://www.designcouncil.org.uk/futurecurrents

90. Brenda Boardman *et al.*, op. cit. (note 23, above), p. 84.

91. George Marshall, Climate Outreach Information Network, personal communication.

92. House of Lords Select Committee on Science and Technology, op. cit. (note 15, above), paras 6.33–6.44.

5 KEEPING THE LIGHTS ON

1. *Faust*, Lines 10216–19

2. PB Power, *The Cost of Generating Electricity* (The Royal Academy of Engineering, London, 2004), p. 13.

3. Godfrey Boyle, Bob Everett and Janet Ramage (eds.), *Energy Systems and Sustainability* (Oxford University Press, Oxford, 2003), p. 379.

4. The Sizewell B nuclear power station has a capacity of 1320 MW. David Milborrow, 'The Practicalities of Developing Renewable Energy', 10 October 2003: http://www.bwea.com/pdf/DM-Lords-intermittency.pdf

5. The Sustainable Development Commission, *Wind Power in the UK*, May 2005, p. 27: http://www.sd-commission.org.uk/publications/downloads/Wind_Energy-Nov Rev2005.pdf

6. The Royal Commission on Environmental Pollution, *Energy – The Changing Climate*, June 2000, Chapter 8, Box 8B: http://www.rcep.org.uk/newenergy.htm

7. The British Wind Energy Association, 'Power UK: Costs and Benefits of Large-scale Development of Wind-power', March 2003:
http://www.bwea.com/pdf/PowerUK-March2003-page17–25.pdf

8. Brenda Boardman *et al.*, *40% House* (Environmental Change Institute, University of Oxford, 2005), p. 74.

9. Department of Trade and Industry, *Energy Trends*, 2005, p. 15:
http://www.dti.gov.uk/files/file11881.pdf

10. The Royal Commission on Environmental Pollution, op. cit. (note 6, above), para 3.36.

11. ibid., para 3.38.

12. Bennett Daviss, 'Coal Goes for the Burn', *New Scientist*, 3 September 2005.

13. Greenpeace UK, 'Decentralising Power: An Energy Revolution for the 21st Century', 2005, p. 26:
http://www.greenpeace.org.uk/MultimediaFiles/Live/FullReport/7154.pdf

14. US Energy Information Administration, 'Country Analysis Brief: United Kingdom', April 2005:
http://www.eia.doe.gov/emeu/cabs/United_Kingdom/Background.html

15. Bennett Davis, op. cit. (note 12, above).

16. ibid.

17. The Department of Trade and Industry says that 'Encouraged by new lower-cost and higher-efficiency technology, combined with an increasing price advantage over natural gas, coal could show significant expansion in UK power generation after 2020,' Department of Trade and Industry, 'A Strategy for Developing Carbon Abatement Technologies for Fossil Fuel Use', 2005, p. 20:
http://www.dti.gov.uk/energy/coal/cfft/catstrategy.shtml

18. Robert L. Hirsch, Roger Bezdek and Robert Wendling, 'Peaking Of World Oil Production: Impacts, Mitigation, and Risk Management', US Department of Energy, February 2005, p. 34, available at
http://www.hubbertpeak.com/us/NETL/OilPeaking.pdf

19. Christopher Adams, 'Industry Feels Heat of Gas Price Surge', *Financial Times*, 17 February 2006.

20. The Economist Intelligence Unit, 'Ukraine Economy: Gas Trouble', 17 February 2006: http://www.tmcnet.com/usubmit/2006/02/17/1384897.htm

21. For example, Martin Flanagan, 'Call for Inquiry as Gas Pipe from Europe Runs Half Empty', *Scotsman*, 30 January 2006.

22. Parliamentary Office of Science and Technology, 'The Future of UK Gas Supplies', Postnote no. 230, October 2004:
http://www.parliament.uk/documents/upload/POSTpn230.pdf

23. The Royal Commission on Environmental Pollution, op. cit. (note 6, above), para 3.36.

24. George Monbiot, letter to *New Scientist*, 11 June 2005.

25. Global Business Environment, Shell International, 2002. 'People and Connections: Global Scenarios to 2020', 2002:

http://www.shell.com/static/media-en/downloads/peopleandconnections.pdf

26. The Geological Society, 'How to Plug the Energy Gap', 10 November 2005:
http://www.geolsoc.org.uk/template.cfm?name=PR60

27. Olav Kaarstad, Statoil, quoted by Reuters, 'Norway Has Vast, Inaccessible Seabed Coal – Statoil', 21 December 2005.

28. The Geological Society, op. cit. (note 26, above).

29. Olav Kaarstad, op. cit. (note 27, above).

30. Bennett Daviss, op. cit. (note 12, above).

31. House of Lords Science and Technology Committee, 'Renewable Energy: Practicalities', 15 July 2004, para 2.5:
http://www.publications.parliament.uk/pa/ld200304/ldselect/ldsctech/126/12602.htm

32. Parliamentary Office of Science and Technology, op. cit. (note 22, above).

33. Dominion, 'Underground Storage', 2006:
http://www.dom.com/about/gas-transmission/storage.jsp

34. Columbia Gas Transmission Corporation, Docket No. CP05–144-000, 2001:
http://www.columbiagastrans.com/NewProjects/Hardy/Virginia%20Looping.htm

35. Performance and Innovation Unit, 'The Energy Review', February 2002, Annex 6: http://www.number-10.gov.uk/su/energy/20.html

36. Department of Trade and Industry, 'A Strategy for Developing Carbon Abatement Technologies for Fossil Fuel Use', 2005:
http://www.dti.gov.uk/energy/coal/cfft/catstrategy.shtml

37. Fred Pearce, 'Squeaky Clean Fossil Fuels', *New Scientist*, 30 April 2005.

38. Department of Trade and Industry, op. cit. (note 36, above), p. 26.

39. Emma Young, 'Burying the Problem', *New Scientist*, 3 September 2005.

40. Department of Trade and Industry, op. cit. (note 36, above).

41. Godfrey Boyle, Bob Everett and Janet Ramage (eds.), op. cit. (note 3, above), p. 578.

42. Department of Trade and Industry, 'Review of the Feasibility of Carbon Dioxide Capture and Storage in the UK', 2005, p. 13:
http://www.dti.gov.uk/files/file21887.pdf

43. Department of Trade and Industry, op. cit. (note 36, above) p. 28.

44. Bert Metz *et al.* (eds.), 'Carbon Dioxide Capture and Storage: Summary for Policymakers', Intergovernmental Panel on Climate Change, no date, Table SPM.1, p. 9: http://www.ipcc.ch/activity/ccsspm.pdf

45. ibid., p. 13.

46. ibid., p. 35.

47. Cited by the Union of Concerned Scientists, 'Policy Context Of Geologic Carbon Sequestration', 2003:
http://www.ucsusa.org/assets/documents/global_warming/GEO_CARBON_SEQ_for_web.pdf

48. Here are some estimates:

source	cost per tonne of buried carbon dioxide
The Cooperative Research Centre for Greenhouse Gas Technologies, present day[49]	US$40 (£23)
UK government's Inter-departmental Analysts Group, estimate for 2020 or 2025[50]	£20–30
Intergovernmental Panel on Climate Change, for gas plants, present day[51]	US$40–90 (£23–52)
OECD/International Energy Agency, present day[52]	US$50–100 (£30–60)
Imperial College MARKAL model, from gas. Estimate for 2020 or 2025[53]	£50–55
Intergovernmental Panel on Climate Change, for coal plants, present day[54]	US$40–270 (£23–160)

49. Cited by Emma Young, op. cit. (note 39, above).

50. Department of Trade and Industry, op. cit. (note 42, above), p. 29.

51. Bert Metz et al. (eds.), op. cit. (note 44, above), p. 17, Table SPM.4.

52. Organization for Economic Cooperation and Development/International Energy Agency, Prospects for CO_2 Capture and Storage (OECD, Paris, December 2004), cited by Department of Trade and Industry, op. cit. (note 42, above).

53. Cited in Department of Trade and Industry, op. cit. (note 42, above), p. 29.

54. Bert Metz et al. (eds.), op. cit. (note 44, above), p. 17, Table SPM.4 (this includes both pulverized coal and Integrated Gasification Combined Cycle).

55. Department of Trade and Industry, 'Energy – Its Impact on the Environment and Society', 2005, p. 43, Table 4 (47MtC, multiplied by 3.667): http://www.dti.gov.uk/files/file20263.pdf

56. Bert Metz et al. (eds.), op. cit. (note 44, above), p. 10.

57. E-mail from Jonathan Gibbins, 27 September 2005.

58. Department of Trade and Industry, op. cit. (note 42, above).

59. Advanced Resources International, 'Undeveloped Domestic Oil Resources: The Foundation for Increasing Oil Production and a Viable Domestic Oil Industry', US Department of Energy, February 2006: http://www.fossil.energy.gov/programs/oilgas/publications/eor_co2/Undeveloped_Oil_Document.pdf

60. Paul Harris, 'They Flattened this Mountaintop to Find Coal – and Created a Wasteland', Observer, 16 January 2005.

61. See, for example, http://www.ohvec.org/galleries/mountaintop_removal/007/43.html

62. Department of Trade and Industry, 'Review of the Feasibility of Underground Coal Gasification in the UK', October 2004, Executive Summary, p. 3: http://www.dti.gov.uk/energy/coal/cfft/ucgfeasibilityreport.pdf

63. ibid., p. 5.

64. ibid., p. 7.

65. US Energy Information Administration, op. cit. (note 14, above).

66. Department of Trade and Industry, op. cit. (note 62, above), p. 3.

67. The Royal Commission on Environmental Pollution, op. cit. (note 6, above), para 8.31.

68. Geoff Dutton *et al.*, *The Hydrogen Energy Economy: Its Long-term Role in Greenhouse Gas Reduction*, Tyndall Centre Technical Report No. 18, January 2005, p. 42.

69. Department of Trade and Industry, op. cit. (note 42, above), p. 33.

70. Organization for Economic Cooperation and Development/International Energy Agency, op. cit. (note 52, above), summarized by Department of Trade and Industry, op. cit. (note 42, above), p. 32.

71. New Economics Foundation, 'Mirage and Oasis: Energy Choices in an Age of Global Warming', 29 June 2005:
http://www.neweconomics.org/gen/uploads/sewyo355prhbgunpscr51d2w2906200 5080838.pdf

72. Greenpeace UK, op. cit. (note 13, above).

73. The Sustainable Development Commission, op. cit. (note 5, above).

74. These states, with the exception of Israel, are listed by Paul Leventhal in an article by William J. Broad, 'Nuclear Weapons in Iran: Plowshare or Sword?', *The New York Times*, 25 May 2004.

75. Israel's weapons programme, as Mordechai Vanunu showed, was developed at the Israeli Atomic Energy Commission site at Dimona, home to one of its two nuclear power plants.

76. 'EC Court Challenge to Sellafield', *BBC News Online*, 3 September 2004:
http://news.bbc.co.uk/1/hi/england/cumbria/3623312.stm

77. Rob Edwards, 'Uranium Pond at Sellafield Sparks Court Threat by EU', *Sunday Herald*, 28 March 2004.

78. 'Legal Threat Over Sellafield Leak', *BBC News Online*, 12 June 2005:
http://news.bbc.co.uk/1/hi/england/cumbria/4085224.stm

79. David Ross, 'Dounreay Admits Shaft Error', *Herald*, 22 May 1997.

80. John Arlidge, 'Fresh Scare on Nuclear Waste', *Guardian*, 2 February 1998.

81. There's an interesting discussion of the conflicting estimates in Godfrey Boyle, Bob Everett and Janet Ramage (eds.), op. cit. (note 3, above), p. 440.

82. Fraser King *et al.*, 'Copper Corrosion under Expected Conditions in a Deep Geologic Repository', Posiva Oy, Helsinki, January 2002:
http://www.posiva.fi/raportit/POSIVA-2002–01.pdf

83. ibid.

84. Rob Edwards, 'Politics Left UK Nuclear Waste Plans in Disarray', *New Scientist*, 18 June 2005.

85. Geoffrey Lean, 'Nuclear Waste: The 1,000-year Fudge', *Independent on Sunday*, 12 June 2005.

86. Erica Werner, 'Yucca Scientists Faked Records to Show Proof of Work', *Pahrump Valley Times*, 6 April 2005:
http://www.pahrumpvalleytimes.com/2005/04/06/news/ymprecords.html

87. anon., 'Pahrump Man in the Center of Yucca Dispute', *Pahrump Valley Times*, 6 April 2005:

http://www.pahrumpvalleytimes.com/2005/04/06/news/ymprecords.html

88. David Lowenthal, University College London, letter to *New Scientist*, 9 July 2005.

89. Nuclear Decommissioning Authority, 'Approved Strategy For Clean-Up of UK's Nuclear Sites Published', press release, 30 March 2006:

http://www.nda.gov.uk/documents/news_release_-_national.pdf

90. Tim Webb and Robin Buckley, 'British Energy has Come Back from the Brink, but How Long Will the Nuclear Frisson Last?', *Independent on Sunday*, 20 June 2004.

91. Paul Brown, 'Taxpayers' £184m Aid to Private Energy Firm', *Guardian*, 18 July 2005.

92. Bridget Woodman, 'New Nuclear Power Plants "Would Halt Move Towards Decentralized Renewable Sources"', *Energy World*, November/December 2005.

93. European Environment Agency, *Energy Subsidies in the European Union: A Brief Overview* (EEA, Copenhagen, 2004), p. 17:

http://reports.eea.eu.int/technical_report_2004_1/en/Energy_FINAL_web.pdf

94. New Economics Foundation, op. cit. (note 71, above), p. 36.

95. ibid., p. 30.

96. ibid.

97. European Environment Agency, op. cit. (note 93, above).

98. Oxera, 'Financing the Nuclear Option: Modelling the Costs of New Build', June 2005, p. 4:

http://www.oxera.com/cmsDocuments/Agenda_June%2005/Financing%20the%20nuclear%20option.pdf

99. Marshall Goldberg, 'Federal Energy Subsidies: Not All Technologies are Created Equal', Renewable Energy Policy Project, July 2000:

http://www.crest.org/repp_pubs/pdf/subsidies.pdf

100. Dwight D. Eisenhower, 'Atoms for Peace', speech to the United Nations General Assembly, 8 December 1953.

101. I'm referring in particular to the Private Finance Initiative schemes I investigated in *Captive State: The Corporate Takeover of Britain* (Macmillan, London, 2000).

102. Ofgem, personal communication.

103. Nuclear Energy Institute, 'Nuclear Power Plants Maintain Lowest Production Cost for Baseload Electricity', 3 September 2003:

http://www.nei.org/index.asp?catnum=4&catid=511

104. PB Power, op. cit. (note 2, above), p. 9.

105. Cited by Performance and Innovation Unit, *The Energy Review*, February 2002, para 42.

106. Performance and Innovation Unit, op. cit. (note 105, above), summary table.

107. Cited by Amory Lovins, 'Nuclear Power: Economics and Climate-protection Potential', Rocky Mountain Institute, 11 September 2005:

http://www.rmi.org/images/other/Energy/E05-08_NukePwrEcon.pdf

108. New Economics Foundation, op. cit. (note 71, above), p. 37.

109. British Energy, 'Sizewell B Power Station Marks 10 Years', 14 February 2005: http://www.british-energy.com/article.php?article=23

110. Godfrey Boyle, Bob Everett and Janet Ramage (eds.), op. cit. (note 3, above), pp. 426–7.

111. World Nuclear Association, 'Uranium availability', 2003, Part 3.3: http://www.world-nuclear.org/education/ne/ne3.htm3.3

112. Jan van Leeuwen and Philip Smith, *Nuclear Power: The Energy Balance*, 6 August 2005 (sixth revision), Chapter 2: http://www.stormsmith.nl/Chap_2_Energy_Production_and_Fuel_costs_rev6.PDF

113. Felix Preston and Paul Baruya, 'Paper 8: Uranium Resource Availability', *The Role of Nuclear Power in a Low Carbon Economy* (Sustainable Development Commission, March 2006): http://www.sd-commission.org.uk/publications/downloads/ Nuclear-paper8-UraniumResourceAvailability.pdf

114. The Sustainable Development Commission, *The Role of Nuclear Power in a Low Carbon Economy*, SDC position paper, March 2006, p. 10: http://www.sd-commission.org.uk/publications/downloads/SDC-Nuclear Position-2006.pdf

115. ibid.

116. Sustainable Development Commission, March 2006. 'Paper 2: Reducing CO_2 Emissions – Nuclear and the Alternatives', *The Role of Nuclear Power in a Low Carbon Economy* (Sustainable Development Commission, March 2006): http://www.sd-commission.org.uk/publications/downloads/ Nuclear-paper2-reducingCO2emissions.pdf

117. World Nuclear Association, 'Energy Balances and CO_2 Implications', November 2005: http://www.world-nuclear.org/info/inf100.htm

118. ibid.

119. Jan van Leeuwen and Philip Smith, op. cit. (note 112, above).

120. http://www.world-nuclear.org/info/inf11.htm

121. http://www.stormsmith.nl/References.PDF

122. Tom Burke, oral evidence to the House of Commons Environmental Audit Committee, 'Keeping the Lights On: Nuclear, Renewables and Climate Change', 19 October 2005, Q69: http://www.publications.parliament.uk/pa/cm200506/cmselect/cmenvaud/uc584-i/ uc58402.htm

123. Cited by New Economics Foundation, op. cit. (note 71, above), p. 27.

124. E-mail from Jennifer McGregor, Scottish and Southern Energy Group, 2nd November 2005.

125. Nick Jenkins, at the 'Decarbonising the UK' conference, 21 September 2005, Church House, Westminster.

126. Jack Doyle, *Taken for a Ride: Detroit's Big Three and the Politics of Pollution* (Four Walls, Eight Windows, New York, 2000), pp. 1–2.

127. Geoff Dutton *et al.*, op. cit. (note 68, above), p. 42.

6 HOW MUCH ENERGY CAN RENEWABLES SUPPLY?

1. *Faust*, lines 259–62.

2. The renewable technology whose manufacture uses the most energy is solar photovoltaic. But even this has an energy payback time of just 2.5–4 years. Erik Alsema and Evert Nieuwlaar, 'Energy Viability of Photovoltaic Systems', *Energy Policy*, Vol. 28 (2002), pp. 999–1010.

3. The House of Lords gives an energy payback time for wind power of 1.1 years. This includes installation and the connection to the grid. House of Lords Science and Technology Committee, 'Renewable Energy: Practicalities', 15 July 2004, Appendix 8, p. 105:
http://www.publications.parliament.uk/pa/ld200304/ldselect/ldsctech/126/12602.htm

4. US Energy Information Administration, 'Country Analysis Brief: United Kingdom', April 2005:
http://www.eia.doe.gov/emeu/cabs/uk.html

5. Energy Technology Support Unit, *New and Renewable Energy: Prospects in the UK for the 21st Century: Supporting Analysis* (ETSU, Harwell, 1999):
http://www.dti.gov.uk/energy/renewables/ publications/pdfs/support.pdf

6. Department of Trade and Industry, 'Offshore Renewables – the Potential Resource', 2005:
http://www.dti.gov.uk/energy/leg_and_reg/consents/future_offshore/chp2.pdf

7. Robert Gross, personal communication.

8. Energy Technology Support Unit, op. cit. (note 5, above), p. 171.

9. Department of Trade and Industry, op. cit. (note 6, above), p. 22.

10. Roberto Rudervall, J. P. Charpentier and Raghuveer Sharma, 'High Voltage Direct Current (HVDC) Transmission Systems: Technology Review Paper', World Bank and ABB, no date, p. 6:
http://www.worldbank.org/html/fpd/em/transmission/technology_abb.pdf

11. ibid.

12. ibid., p. 18.

13. ABB Group, 'History of ABB's HVDC expertise', 2005:
http://www.abb.ee/global/abbzh/abbzh251.nsf!OpenDatabase&db=/global/abbzh/abbzh250.nsf&v=553E&e=us&url=/global/seitp/seitp202.nsf/0/7CFD9A3A7416A383C1256E8600406F4F!OpenDocument

14. David Milborrow, 'Offshore Wind Rises to the Challenge', *Wind Power Monthly*, April 2003, p. 51, cited by Robert Gross, 'Technologies and Innovation for System Change in the UK: Status, Prospects and System Requirements of Some Leading Renewable Energy Options', *Energy Policy*, Vol. 32 (2004), pp. 1905–19.

15. ibid.

16. Robert Gross, 'Technologies and innovation for system change in the UK: status, prospects and system requirements of some leading renewable energy options', *Energy Policy*. Vol. 32 (2004), pp. 1905–19.

17. Greenpeace, Estia and Solar Paces, 'Concentrated Solar Thermal Power – Now', September 2005:

http://www.solarpaces.org/051006%20Greenpeace-Concentrated-Solar-Thermal-Po
wer-Now-2005.pdf
18. ibid., p. 13.
19. ibid., p. 27.
20. Fred Pearce, 'Power of the Midday Sun', *New Scientist*, 10 April 2004.
21. Greenpeace, Estia and Solar Paces, op. cit. (note 17, above).
22. Rachel Nowak, 'Power Tower', *New Scientist*, 31 July 2004.
23. Kosuke Kurokawa (ed.), 'Energy from the Desert – Feasibility of Very Large
Scale Photovoltaic Power (VLS-PV) Systems', IEA Task VIII (James and James Ltd,
May 2003), summarized at:
http://jxj.base10.ws/magsandj/rew/2003_03/desert_power.html
24. International Energy Agency, *Electricity Information* (IEA, Paris, 2005), Part I
– 1.3.
25. Kosuke Kurokawa (ed.), op. cit. (note 23, above).
26. ibid.
27. The Sustainable Development Commission, 'Wind Power in the UK', May 2005,
p. 20:
http://www.sd-commission.org.uk/publications/downloads/Wind_Energy-NovRev
2005.pdf
28. National Grid Company, cited by the Sustainable Development Commission,
op. cit. (note 27, above), p. 23.
29. ibid.
30. See the flattening curve on page 31 of Goran Strbac and Ilex Energy Consulting,
'Quantifying the System Costs of Additional Renewables in 2020: A Report to the
Department of Trade and Industry', October 2002:
http://www.dti.gov.uk/energy/developep/080scar_report_v2_0.pdf
31. Graham Sinden, 'Diversified Renewable Energy Portfolios for the UK', presentation to the British Institute of Energy Economics 'Diversified Renewable Strategy'
conference, 22 September 2005. In this report, he estimates a capacity credit of
8GW, but has since revised it to 6GW (personal communication).
32. Goran Strbac and Ilex Energy Consulting, op. cit. (note 30, above), Table 1,
page ii: http://www.dti.gov.uk/energy/developep/080scar_report_v2_0.pdf
33. Robert Gross *et al.*, 'The Costs and Impacts of Intermittency: An Assessment of
the Evidence on the Costs and Impacts of Intermittent Generation on the British
Electricity Network', UK Energy Research Centre, April 2006.
34. The Royal Commission on Environmental Pollution, *Energy – The Changing
Climate*, June 2000, paragraph 8.54: http://www.rcep.org.uk/newenergy.htm
35. Robert Gross, op. cit. (note 16, above).
36. Goran Strbac and Ilex Energy Consulting, op. cit. (note 32, above).
37. The Royal Commission on Environmental Pollution, op. cit. (note 34, above),
para 8.41.
38. Lewis Dale *et al.*, 'Total Cost Estimates for Large-scale Wind Scenarios in the
UK', *Energy Policy*, Vol. 32 (2004), pp. 1949–56.
39. Hugh Sharman, 'Why UK Wind Power Should Not Exceed 10 GW', *Civil
Engineering*, Vol. 158 (November 2005), pp. 161–9.

40. ibid.

41. Ofgem, personal communication.

42. Performance and Innovation Unit, *The Energy Review*, February 2002, Annex 6, para 40: http://www.number-10.gov.uk/su/energy/20.html

43. Cited by New Economics Foundation, 'Mirage and Oasis: Energy Choices in an Age of Global Warming', 29 June 2005: http://www.neweconomics.org/gen/uploads/sewyo355prhbgunpscr51d2w2906200 5080838.pdf

44. Performance and Innovation Unit, op. cit. (note 42, above), para 38.

45. The Sustainable Development Commission, op. cit. (note 27, above), p. 27.

46. PB Power, *The Cost of Generating Electricity* (The Royal Academy of Engineering, London, 2004), pp. 45, 47.

47. ibid.

48. Performance and Innovation Unit, op. cit. (note 42, above), paras 55, 61.

49. The Sustainable Development Commission, op. cit. (note 27, above).

50. Performance and Innovation Unit, op. cit. (note 42, above), summary of key findings.

51. Jake Chapman and Robert Gross, 'Technical and Economic Potential of Renewable Energy Generating Technologies: Potentials and Cost Reductions to 2020', 2001: http://www.strategy.gov.uk/downloads/files/PIUh.pdf

52. ibid.

53. Performance and Innovation Unit, op. cit. (note 42, above), para 40.

54. PB Power, op. cit. (note 46, above), p. 49.

55. ibid.

56. Dave Andrews, Wessex Water, 'How Diesels are Used to Provide Standing Reserve Services to the National Grid', presentation to the Open University 'Wind Power and Renewables' conference, 11 January 2006.

57. Dave Andrews, Wessex Water, 'The Availability Of Rapid Start Standby Generation And Its Potential Role in Coping With Intermittency Due to a Large Penetration of Renewables', unpublished note, 4 July 2005.

58. Godfrey Boyle (ed.), *Renewable Energy: Power for a Sustainable Future* (Oxford University Press, Oxford, 2004), p. 405.

59. Graham Sinden, op. cit. (note 31, above).

60. Dave Andrews, op. cit. (note 56, above).

61. ibid.

62. Oxera, 'The Non-Market Value of Generation Technologies', June 2003, p. 14: http://www.oxera.co.uk/oxera/publicnsf/images/DKIG-5NCERA/$file/OXERA Report.pdf

63. Graham Sinden, 'Assessing the Costs of Intermittent Power Generation', presentation to the Intermittency Stakeholder Group, 5 July 2005.

64. ibid.

65. Graham Sinden, op. cit. (note 31, above).

66. For example, The Royal Commission on Environmental Pollution, op. cit. (note 34, above), para 8.57.

67. Compound figures adjusted for inflation. Compiled for Dave Andrews, personal communication.
68. House of Lords Science and Technology Committee, op. cit. (note 3, above), para 7.21:
http://www.publications.parliament.uk/pa/ld200304/ldselect/ldsctech/126/12602.htm
69. The Royal Commission on Environmental Pollution, op. cit. (note 34, above), para 8.58.
70. Brenda Boardman *et al.*, *40% House* (Environmental Change Institute, University of Oxford, 2005), p. 39.
71. Gaia Vince, 'Smart Fridges Could Ease Burden on Energy Supply', *New Scientist*, 30 July 2005.
72. Robert Gross, op. cit. (note 16, above).
73. UK Energy Research Centre, 'An Assessment of Evidence on the Costs of Intermittent Power Generation: What is the Evidence on the Costs and Engineering Impacts of Intermittent Generation on the UK Electricity Network, and how Those Costs are Assigned?', summary note, Stakeholder workshop, Imperial College, 5 July 2005.
74. Oliver Tickell, 'Real-time Pricing Initiative', unpublished paper, 2005.
75. ibid.
76. Suleiman Abu-Sharkh *et al.*, *Microgrids: Distributed On-site Generation*, Tyndall Centre Technical Report No. 22, March 2005, p. 61.
77. Brenda Boardman *et al.*, op. cit. (note 70, above), p. 81.
78. Jeremy Woods, Robert Gross and Matthew Leach, Centre for Energy Policy and Technology, *Innovation in the Renewable Heat Sector in the UK: Markets, Opportunities and Barriers*, Imperial College, London, December 2003, p. 4:
http://www.dti.gov.uk/renewables/policy/iceptinnovationbarriers.pdf
79. ibid., p. 5.
80. ibid.
81. Jake Chapman and Robert Gross, op. cit. (note 51, above).
82. Royal Commission on Environmental Pollution, 'Biomass as a Renewable Energy Source', 2004: http://www.rcep.org.uk/biomass/Biomass%20Report.pdf
83. UN Food and Agriculture Organization, 'Parameters, Units and Conversion Factors', no date:
http://www.fao.org/documents/show_cdr.asp?url_file=/docrep/007/j4504e/j4504e08.htm
84. The Royal Commission on Environmental Pollution, op. cit. (note 34, above), para 7.76.
85. Fred Pearce, *When the Rivers Run Dry* (Eden Project Books, London, 2006).
86. Cited by Lord Whitty, Minister for Food, Farming and Sustainable Energy, Supplementary Memorandum, 17 September 2003:
http://www.publications.parliament.uk/pa/cm200203/cmselect/cmenvfru/929/3091509.htm
87. Energy Technology Support Unit, op. cit. (note 5, above), p. 73.
88. Royal Commission on Environmental Pollution, op. cit. (note 34, above), para 3.47.

89. House of Lords Science and Technology Committee, op. cit. (note 68, above), para 4.26.

90. Lord Sainsbury of Turville, parliamentary answer, 22 March 2004: http://www.publications.parliament.uk/pa/ld200304/ldhansrd/vo040322/text/40322w03.htm

91. Royal Commission on Environmental Pollution, op. cit. (note 82, above), para 2.76.

92. Energy Technology Support Unit, op. cit. (note 5, above), Table 8, p. 66.

93. UN Food and Agriculture Organization, op. cit. (note 82, above).

94. Energy Technology Support Unit, op. cit. (note 5, above), p. 66.

95. L. M. Maene, 'Phosphate Fertilizer Production, Consumption and Trade: The Present Situation and Outlook to 2010', paper presented to the Sulphur Institute's 17th Sulphur Phosphate Symposium, Boca Raton, Florida, January 17–19, 1999. L. M. Maene is the Director General of the International Fertilizer Industry Association, Paris. http://www.fertilizer.org/ifa/publicat/PDF/1999_biblio_54.pdf

96. Future Energy Solutions, AEA Technology, 'Renewable Heat and Heat from Combined Heat and Power Plants', April 2005, p. 46: http://www.defra.gov.uk/farm/acu/energy/fes-renewable-chp.pdf

97. ibid., p. 53.

98. Royal Commission on Environmental Pollution, op. cit. (note 34, above), para 3.43.

99. Future Energy Solutions, AEA Technology, op. cit. (note 96, above), p. 42.

100. New Economics Foundation, op. cit. (note 43, above), p. 23.

101. Future Energy Solutions, AEA Technology, op. cit. (note 96, above), p. 60.

7 THE ENERGY INTERNET

1. *Faust*, lines 11072–4.

2. Walt Patterson, 'Keeping the Lights On', Working Papers 1–3, The Royal Institute for International Affairs, March 2003: http://www.chathamhouse.org.uk/index.php?id=233

3. Suleiman Abu-Sharkh *et al.*, *Microgrids: Distributed On-site Generation*, Tyndall Centre Technical Report No. 22, March 2005.

4. Greenpeace UK, 'Decentralising Power: An Energy Revolution for the 21st Century', 2005: http://www.greenpeace.org.uk/MultimediaFiles/Live/FullReport/7154.pdf

5. Erik Alsema and Evert Nieuwlaar, 'Energy Viability of Photovoltaic Systems', *Energy Policy*, Vol. 28 (2000), pp. 999–1010.

6. Jeremy Leggett, *Half Gone: Oil, Gas, Hot Air and the Global Energy Crisis* (Portobello Books, London, 2005), p. 201.

7. ibid., note 253, p. 290.

8. Energy Technology Support Unit, *New and Renewable Energy: Prospects in the UK for the 21st Century – Supporting Analysis* (ETSU, Harwell, 1999), p. 141.

9. ibid., Figure 1: PV Resource-Cost Curve, p. 143.

10. Suleiman Abu-Sharkh *et al.*, op. cit. (note 3, above), p. 33.

11. Jake Chapman and Robert Gross, 'Technical and Economic Potential of Renewable Energy Generating Technologies: Potentials and Cost Reductions to 2020', 2001: http://www.strategy.gov.uk/downloads/files/PIUh.pdf

12. Suleiman Abu-Sharkh *et al.*, op. cit. (note 3, above), p. 53.

13. Robert Gross, 'Technologies and Innovation for System Change in the UK: Status, Prospects and System Requirements of Some Leading Renewable Energy Options', *Energy Policy*, Vol. 32 (2004), pp. 1905–19.

14. http://www.etsap.org/markal/main.html

15. Department of Trade and Industry, Energy White Paper, 2003, Supplementary Annexes, p. 7:
http://www.dti.gov.uk/energy/whitepaper/annexes.pdf

16. Ofgem, personal communication.

17. Performance and Innovation Unit, *The Energy Review*, February 2002, Annex 6, para 40: http://www.number-10.gov.uk/su/energy/20.html

18. Jeremy Leggett, 'Here Comes the Sun', *New Scientist*, 6 September 2003.

19. The more pessimistic figure (thirty-five years) comes from Jim Watson, 'Co-Provision in Sustainable Energy Systems: The Case of Microgeneration', *Energy Policy*, Vol. 32 (March 2004), pp. 1981–90.

20. Energy Technology Support Unit, op. cit. (note 8, above), p. 133.

21. Royal Commission on Environmental Pollution, *Energy – The Changing Climate*, June 2003, Para 7.40: http://www.rcep.org.uk/newenergy.htm

22. Godfrey Boyle (ed.), *Renewable Energy: Power for a Sustainable Future* (Oxford University Press, Oxford, 2004), pp. 76–82.

23. Jake Chapman and Robert Gross, op. cit. (note 11, above).

24. Paul Gipe, quoted by Nick Martin, 'Can We Harvest Useful Wind Energy from the Roofs of Our Buildings?', *Building for a Future*, winter 2005/6, special wind-power feature, Table 2.

25. Nick Martin, op. cit. (note 24, above), Table 2.

26. ibid.

27. ibid.

28. Derek Taylor, 'Potential Outputs from 1–2m Diameter Wind Turbines', *Building for a Future*, winter 2005/6, special wind-power feature, Figure 1.

29. Oliver Lowenstein, 'Green Towers', *Building for a Future*, June 2005.

30. Godfrey Boyle (ed.), op. cit. (note 22, above), p. 284.

31. Chris Dunham, personal communication.

32. Mick Hamer, 'Be a Power Broker in Your Own Home', *New Scientist*, 14 February 2004.

33. Fred Pearce, 'Generate and Sell Your Own Electricity', *New Scientist*, 21 June 2003.

34. Department of Trade and Industry, Energy White Paper: 'Our Energy Future – Creating a Low Carbon Economy', 2003, para 4.16.

35. Mick Hamer, op. cit. (note 32, above).

36. Future Energy Solutions, AEA Technology, 'Renewable Heat and Heat from Combined Heat and Power Plants', April 2005, p. 19:

http://www.defra.gov.uk/farm/acu/energy/fes-renewable-chp.pdf

37. Suleiman Abu-Sharkh *et al.*, op. cit. (note 3, above), p. 70.

38. Neil Crumpton, personal communication.

39. ibid.

40. http://www.statistics.gov.uk/census2001/profiles/printV/UK.asp

41. Department of Trade and Industry, *Digest of UK Energy Statistics*, 2005: http://www.dti.gov.uk/energy/inform/dukes/

42. Fred Pearce, op. cit. (note 33, above).

43. Future Energy Solutions, AEA Technology, op. cit. (note 36, above), p. 18.

44. Energy Saving Trust, Econnect and Element Energy, 'Potential for Microgeneration: Final Report', 14 November 2005: http://portal.est.org.uk/uploads/documents/aboutest/Microgeneration%20in%20 the%20UK%20-%20final%20report%20REVISED_executive%20summary1.pdf

45. Godfrey Boyle, Bob Everett and Janet Ramage (eds.) *Energy Systems and Sustainability*, Oxford University Press, Oxford, 2003, p. 371.

46. Future Energy Solutions, AEA Technology, op. cit. (note 37, above).

47. House of Lords Select Committee on Science and Technology, 'Energy Efficiency', 2005, Appendix 7 (Visit to Germany): http://www.parliament.the-stationery-office.co.uk/pa/ld200506/ldselect/ldsctech/ 21/2122.htm

48. Future Energy Solutions, AEA Technology, op. cit. (note 36, above).

49. Geoff Dutton *et al.*, *The Hydrogen Energy Economy: Its Long-term Role in Greenhouse Gas Reduction*, Tyndall Centre Technical Report No. 18, January 2005, p. 12.

50. National Academy of Engineering, 'The Hydrogen Economy: Opportunities, Costs, Barriers, and R&D Needs', 2004, Table 4–1: http://www.nap.edu/books/0309091632/html/39.html

51. National Academy of Engineering, op. cit. (note 49, above), Table 4–1.

52. ibid.

53. Centre for Energy Policy and Technology, 'Assessment of Technological Options to Address Climate Change', Imperial College, London, 2002, cited by Godfrey Boyle (ed.), op. cit. (note 22, above), p. 410.

54. ibid.

55. Ofgem, personal communication.

56. Godfrey Boyle, Bob Everett and Janet Ramage (eds)., op. cit. (note 45, above), p. 591.

57. Geoff Dutton *et al.*, op. cit. (note 50, above), p. 13.

58. ibid., p. 14.

59. National Academy of Engineering, op. cit. (note 51, above), p. 38: http://darwin.nap.edu/books/0309091632/html/38.html

60. Zhili Feng et al., 'V.A. Pipelines: V.A.1 Hydrogen Permeability and Integrity of Hydrogen Transfer Pipelines', US Department of Energy, 2005: http://www.hydrogen.energy.gov/pdfs/progress05/v_a_1_feng.pdf

61. National Academy of Engineering, op. cit. (note 51, above), p. 38: http://darwin.nap.edu/books/0309091632/html/38.html

62. National Academy of Engineering, op. cit. (note 51, above), p. 39:
http://darwin.nap.edu/books/0309091632/html/38.html

63. Royal Commission on Environmental Pollution, op. cit. (note 21, above), para 3.42.

64. National Academy of Engineering, op. cit. (note 51, above), p. 31:
http://www.nap.edu/openbook/0309091632/html/31.html

65. *Decarbonising the UK – Energy for a Climate Conscious Future* (The Tyndall Centre for Climate Change Research, 2005), p. 45:
http://www.tyndall.ac.uk/media/news/tyndall_decarbonising_the_uk.pdf

66. Royal Commission on Environmental Pollution, op. cit. (note 21, above), Box 8D.

67. Anil Ananthaswamy, 'Reality Bites for the Dream of a Hydrogen Economy', *New Scientist*, 15 November 2003.

68. ibid.

69. US Department of Energy, 'President's Hydrogen Fuel Initiative', 2006:
http://www1.eere.energy.gov/hydrogenandfuelcells/presidents_initiative.html

70. See, for example, http://www.bakuceyhan.org.uk/

71. Suleiman Abu-Sharkh *et al.*, op. cit. (note 3, above).

72. Greenpeace UK, op. cit. (note 4, above), p. 5.

73. House of Lords Science and Technology Committee, 'Renewable Energy: Practicalities', 15 July 2004, Appendix 11:
http://www.publications.parliament.uk/pa/ld200304/ldselect/ldsctech/126/12602.htm

74. Suleiman Abu-Sharkh *et al.*, op. cit. (note 3, above), p. 9.

75. House of Lords Select Committee on Science and Technology, op. cit. (note 47, above), para 2.14.

76. Greenpeace UK, op. cit. (note 4, above), p. 28.

77. International Energy Agency, 'World energy investment outlook', 2003, cited by Greenpeace, op. cit. (note 4, above), p. 28.

8 A NEW TRANSPORT SYSTEM

1. *Faust*, line 7118.

2. Ilya Ehrenburg, *The Life of the Automobile* (Serpent's Tail, London, 1999 [1929]), p. 175.

3. George Monbiot, *Amazon Watershed* (Michael Joseph, London, 1991).

4. Stephen Khan, 'Saboteurs Take Out 700 Speed Cameras', *Observer*, 7 September 2003.

5. See the links page at http://www.abd.org.uk/

6. In 1938 Leslie Burgin, the UK Minister of Transport, said 'the experience of my Department is that the construction of a new road tends to result in a great increase in traffic, not only on the new road but also on the old one which it was built to supersede.' Quoted by the Standing Advisory Committee on Trunk Road Assessment,

'Trunk Roads and the Generation of Traffic', Department for Transport, 1994, para 4.10, cited at http://www.cbc.org.nz/Resources/whycars.shtml

7. Quoted by Paul Brown, 'Prescott Points to Buses in Fast Lane', *Guardian*, 6 June 1997.

8. Department for Transport, *Transport Statistics Great Britain, 2005*, October 2005, p. 121:
http://www.dft.gov.uk/stellent/groups/dft_transstats/documents/downloadable/dft_transstats_609987.pdf

9. Department for Transport, *The Future of Transport: Modelling and Analysis*, 2005, p. 10:
http://www.dft.gov.uk/stellent/groups/dft_about/documents/downloadable/dft_about_036814.pdf

10. ibid., Figure 3.6, p. 9.

11. Department for Transport 2005, cited by Transport 2000, 2006 – see note 8.

12. ibid.

13. Department for Transport, op. cit. (note 9, above), p. 6.

14. Department for Transport, 'National Travel Survey 2003: Provisional Results', 2004: http://www.dft.gov.uk/pns/DisplayPN.cgi?pn_id=2004_0103

15. National Travel Survey, cited by Transport 2000, 2006 – see note 8.

16. Department for Transport/Office of National Statistics, 2005, cited by Transport 2000, 2006 – see note 8.

17. Department for Transport, op. cit. (note 8, above), p. 15.

18. Department for Transport, op. cit. (note 9, above), Table A3, p. 16.

19. ibid., p. 11.

20. Sylviane de Saint-Seine, 'Cars Won't make '08 CO_2 Goal', *Automotive News*, 25 July 2005:
http://www.autonews.com/apps/pbcs.dll/article?AID=/20050725/SUB/507250857&SearchID=73238030583873

21. Department for Transport statistics, December 2005, collated by Road Block:
http://www.roadblock.org.uk/press_releases/info/TPI%20and%20local%20schemes%20Dec05.xls

22. David Jamieson, transport minister, parliamentary answer, Hansard column 786W, 8 July 2004:
http://www.parliament.the-stationery-office.co.uk/pa/cm200304/cmhansrd/vo040708/text/407 08w05.htm

23. Freight on Rail and TRANSform Scotland, submission to the Freight Transport Inquiry by the Local Government and Transport Committee, Scottish Parliament, 2 December 2005:
http://www.freightonrail.org.uk/ConsultationsLocalGovernmentandTransport.htm

24. Freight on Rail, 'Useful Facts and Figures', 2005:
http://www.freightonrail.org.uk/FactsFigures.htm

25. Alan Storkey, 'A Motorway-based National Coach System', 2005, available from alan@storkey.com

26. ibid.

27. The Highway Code, cited by Alan Storkey, op. cit. (note 25, above).

28. Alan Storkey, op. cit. (note 25, above).

29. ibid.

30. ibid.

31. ibid.

32. AEA Technology, *CAFE CBA: Baseline Analysis 2000 to 2020*, April 2005, p. 109. It estimates 39,470 deaths caused by chronic exposure to air pollution, and 1,320 by acute exposure:
http://europa.eu.int/comm/environment/air/cafe/activities/pdf/cba_baseline_results 2000_2020.pdf

33. ibid., p. 29.

34. http://www.cia.gov/cia/publications/factbook/rankorder/2119rank.html

35. Alan Storkey, op. cit. (note 25 above).

36. ibid.

37. Royal Academy of Engineering, 'Transport 2050: The Route to Sustainable Wealth Creation', March 2005:
http://www.raeng.org.uk/news/publications/list/reports/Transport_2050.pdf

38. Tim Collins MP, Hansard Column 1180, 9 July 2003:
http://www.publications.parliament.uk/pa/cm200203/cmhansrd/vo030709/debtext/ 30709-09.htm

39. Alistair Darling, Secretary of State for Transport, Hansard Column 1181, 9 July 2003:
http://www.publications.parliament.uk/pa/cm200203/cmhansrd/vo030709/debtext/ 30709-09.htm

40. Robert L. Hirsch, Roger Bezdek and Robert Wendling, 'Peaking Of World Oil Production: Impacts, Mitigation, and Risk Management', US Department of Energy, February 2005, available at http://www.hubbertpeak.com/us/NETL/OilPeaking.pdf

41. HM Treasury, Budget 2006, Chapter 7, p. 17, Chart 7.3:
http://www.hm-treasury.gov.uk/media/20F/1D/bud06_ch7_161.pdf

42. US Environmental Protection Agency, 'Green Vehicle Guide', 2006:
http://www.epa.gov/greenvehicles/all-rank-06.htm

43. Toyota, 'Toyota Displays Earth-friendly ES3 Concept Car at International Frankfurt Motor Show 2001', 12 September 2001:
http://www.toyota.co.jp/en/news/01/0912.html

44. Peugeot advertisement, 'A New Number to be Reckoned With', *Daily Telegraph*, 20 October 1983, p. 9.

45. US Environmental Protection Agency, op. cit. (note 42, above).

46. Sylviane de Saint-Seine, op. cit. (note 20, above).

47. The Society of Motor Manufacturers and Traders, 'UK New Car Registrations by CO_2 Performance', April 2005:
http://lib.smmt.co.uk/articles/sharedfolder/Publications/CO2Report%20New.pdf

48. Jeff Plungis, 'Engineers Push Fuel Economy to Front Seat at Auto Summit', *Detroit News*, 11 April 2005:
http://www.detnews.com/2005/autoinsider/0504/12/A01-146552.htm

49. *Detroit News*, 4 June 2003, cited by Want to Know, 'Car Mileage: 1908 Ford Model T – 25mpg; 2004 EPA Average All Cars – 21mpg', 11 July 2005:
http://www.wanttoknow.info/050711carmileageaveragempg

50. US Environmental Protection Agency, op. cit. (note 42, above), SUVs:
http://www.epa.gov/greenvehicles/suv-06.htm

51. David Johns, quoted in 'Car Efficiency Figures "Misleading" ', *Guardian Unlimited*, 26 October 2005:
http://money.guardian.co.uk/cars/story/0,11944,1600941,00.html

52. Ann Job, 'Blame the Feds for Fuel Economy Figures that Don't Match Real World', *Detroit News*, 19 May 2004.

53. anon., 'Green Machines', *Which?*, May 2006.

54. Paul Hawken, Amory B. Lovins and L. Hunter Lovins, *Natural Capitalism: The Next Industrial Revolution* (Earthscan, London, 1999), pp. 24–32.

55. George W. Bush, 'President Discusses Biodiesel and Alternative Fuel Sources, Virginia BioDiesel Refinery, West Point, Virginia', 16 May 2005:
http://www.whitehouse.gov/news/releases/2005/05/20050516.html

56. White House, FY 2007 Budget, 2006:
http://www.whitehouse.gov/infocus/budget/2007/states/wy.html

57. US House Committee on Energy and Commerce, 'Energy Policy Act 2005: Summary', 2006:
http://energycommerce.house.gov/108/home/Facts_Energy_Policy_Act_2005.pdf

58. 'Directive 2003/30/EC of the European Parliament and of the Council of 8 May 2003 on the Promotion of the Use of Biofuels or Other Renewable Fuels for Transport', *Official Journal of the European Union*, 17 May 2005, Article 3, 1b(ii):
http://ec.europa.eu/energy/res/legislation/doc/biofuels/en_final.pdf

59. ibid., para 17.

60. British Association for Biofuels and Oils, 'Memorandum to the Royal Commission on Environmental Pollution', no date:
http://www.biodiesel.co.uk/press_release/royal_commission_on_environmenta.htm

61. Department for Transport, op. cit. (note 8, above), Table 3.1, p. 50.

62. Department for Environment, Food and Rural Affairs, 'Crops for Energy Branch', 17 November 2004, personal communication.

63. ibid.

64. Department for Environment, Food and Rural Affairs, 'Agriculture in the UK 2003', 2004: http://statistics.defra.gov.uk/esg/publications/auk/2003/chapter3.pdf

65. Lester R. Brown, *The Agricultural Link: How Environmental Deterioration Could Disrupt Economic Progress*, Worldwatch Paper 136 (Worldwatch Institute, Washington, DC, 1997).

66. Friends of the Earth *et al.*, 'The Oil for Ape Scandal: How Palm Oil is Threatening Orang-utan Survival', research report, September 2005, p. 13:
http://www.foe.co.uk/resource/reports/oil_for_ape_full.pdf

67. ibid., p. 10.

68. Peter Aldhous, 'Borneo is Burning', *Nature*, Vol. 432 (11 November 2004), pp. 144–6.

69. C S Tan, 'All Plantation Stocks Rally', *Malaysia Star*, 6 October 2005: http://biz.thestar.com.my/news/story.asp?file=/2005/10/6/business/12243819&sec= business

70. David Bassendine, 'Biofuel: A Self-defeating Policy?', letter to Michael Fallon MP, 8 February 2006: http://www.tonderai.co.uk/index.php/?p=32

71. For example, Tamimi Omar, 'Felda to Set Up Largest Biodiesel Plant', *The Edge Daily*, 1 December 2005:
http://www.theedgedaily.com/cms/content.jsp?id=com.tms.cms.article.Article_ e5d7c0d9-cb73c03a-df4bfc00-d453633e;
Zaidi Isham Ismail, 'IOI to Go It Alone on First Biodiesel Plant', 7 November 2005: http://www.btimes.com.my/Current_News/BT/Monday/Frontpage/20051107000 223/Article/; anon.,
'GHope Nine-month Profit Hits RM841mil', 25 November 2005:
http://biz.thestar.com.my/news/story.asp?file=/2005/11/25/business/12693859& sec=business; anon.,
'GHope to Invest RM40mil for Biodiesel Plant in Netherlands', 26 November 2005: http://biz.thestar.com.my/news/story.asp?file=/2005/11/26/business/12704187& sec=business; anon.,
'Malaysia IOI Eyes Green Energy Expansion in Europe', 23 November 2005: http://www.planetark.com/dailynewsstory.cfm/newsid/33622/story.htm

72. Department for Transport, 'Renewable Transport Fuel Obligation (RTFO) Feasibility Report', November 2005:
http://www.dft.gov.uk/stellent/groups/dft_roads/documents/page/ dft_roads_610329–01.hcspP. 18_263

73. E4Tech, ECCM and Imperial College, London, 'Feasibility Study on Certification for a Renewable Transport Fuel Obligation: Final Report', June 2005.

74. Katharina Kröger, Malcolm Fergusson and Ian Skinner, *Critical Issues in Decarbonising Transport: The Role of Technologies*, Tyndall Centre Working Paper No. 36, October 2003, p. 4:
http://tyndall.e-collaboration.co.uk/publications/working_papers/wp36.pdf

75. anon., 'The Clean Green Energy Dream', *New Scientist Energy Special – Hydrogen*, 16 August 2003.

76. In 2005 George Bush reported, 'We've already dedicated $1.2 billion to hydrogen fuel cell research. I've asked Congress for an additional $500 million over five years to get hydrogen cars into the dealership lot.' George W. Bush, op. cit. (note 55, above).

77. National Academy of Engineering, 2004. 'The Hydrogen Economy: Opportunities, Costs, Barriers, and R&D Needs', 2004, p. 64:
http://darwin.nap.edu/books/0309091632/html/64.html

78. anon., op. cit. (note 75, above).

79. Alison Pridmore and Abigail Bristow, *The Role of Hydrogen in Powering Road Transport*, Tyndall Centre Working Paper No. 19, April 2002, p. 8.

80. Paul Hawken, Amory B. Lovins and L. Hunter Lovins, op. cit. (note 54, above), pp. 26–7.

81. D. R. Blackmore and A. Thomas (eds.), *Fuel Economy of the Gasoline Engine* (Wiley, New York, 1977), p. 42, cited by Byron Wine:
http://byronw.www1host.com

82. National Academy of Engineering, op. cit. (note 77, above), p. 4:
http://www.nap.edu/books/0309091632/html/4.html

83. National Academy of Engineering, op. cit. (note 77, above), p. 27:
http://darwin.nap.edu/books/0309091632/html/27.html

84. National Academy of Engineering, op. cit. (note 77, above), p. 37:
http://darwin.nap.edu/books/0309091632/html/37.html

85. Geoff Dutton *et al.*, *The Hydrogen Energy Economy: Its Long-term Role in Greenhouse Gas Reduction*, Technical Centre Technical Report No. 18, January 2005, p. 79.

86. National Academy of Engineering, op. cit. (note 77, above), p. 119:
http://www.nap.edu/books/0309091632/html/119.html

87. National Academy of Engineering, op. cit. (note 77, above), Table 5.1, p. 46:
http://www.nap.edu/openbook/0309091632/html/46.html

88. Performance and Innovation Unit, *The Energy Review*, February 2002, para 5.21: http://www.strategy.gov.uk/downloads/ su/energy/TheEnergyReview.pdf

89. *Decarbonising the UK – Energy for a Climate Conscious Future* (The Tyndall Centre for Climate Change Research, 2005), p. 77:
http://www.tyndall.ac.uk/media/news/tyndall_decarbonising_the_uk.pdf

90. National Academy of Engineering, op. cit. (note 77, above), p. 83:
http://darwin.nap.edu/books/0309091632/html/83.html

91. anon., op. cit. (note 75, above).

92. Lynn Sloman, *Car Sick: Solutions for Our Car-addicted Culture* (Green Books, Totnes, Devon, 2006).

93. A study in Darlington by Werner Brög, cited by Lynn Sloman, op. cit. (note 92, above), p. 55.

94. Lynn Sloman, op. cit. (note 92, above).

95. Department of the Environment, Transport and the Regions, *A New Deal for Transport: Better for Everyone* (The Stationery Office, London, 1998).

96. S. Cairns *et al.*, *Smarter Choices – Changing the Way We Travel*, Department for Transport, 2004, Chapter 10 (Teleworking):
http://www.dft.gov.uk/stellent/groups/dft_susttravel/documents/page/
dft_susttravel_029753.pdf

97. ibid.

98. ibid.

99. J. Dodgson, J. Pacey and M. Begg, 'Motors and Modems Revisited: The Role of Technology in Reducing Travel Demands and Traffic Congestion', report by NERA for the RAC Foundation and the Motorists Forum, 2000, cited by S. Cairns *et al.*, op. cit. (note 96, above).

100. C. Geraghty, 2004. 'How the Internet Can Help Ease Traffic Congestion', presentation at British Telecom 'Alternative Approaches to Congestion' conference, 26 January 2002, cited by S. Cairns *et al.*, op. cit. (note 96, above).

101. S. Cairns *et al.*, *Making Travel Plans Work: Research Report* (Department for Transport, London, 2002), cited by S. Cairns *et al.*, *Smarter Choices – Changing the Way We Travel*, 2004, Chapter 9 (Car-sharing Schemes): http://www.dft.gov.uk/stellent/groups/dft_susttravel/documents/page/dft_sust travel_029730.pdf

102. ibid.

103. http://www.liftshare.org

104. http://www.liftshare.org/stats.asp

9 LOVE MILES

1. *Faust*, lines 10041–2.

2. E-mail sent by responsibletravel.com, 11 October 2005.

3. http://www.responsibletravel.com/copy/copy100427.htm

4. http://www.responsibletravel.com/Trip/Trip900467.htm

5. http://www.responsibletravel.com/Trip/Trip900445.htm

6. http://www.responsibletravel.com/Trip/Trip100219.htm

7. http://www.responsibletravel.com/TripSearch/Ecotourism/ActivityCategory 100020.htm

8. See George Monbiot, *No Man's Land: An Investigative Journey through Kenya and Tanzania* (Macmillan, London, 1994; republished by Green Books, Totnes, Devon, 2003).

9. Anita Roddick, 'Travel that Doesn't Cost the Earth', *Independent*, 20 September 2005.

10. Royal Commission on Environmental Pollution, 'The Environmental Effects of Civil Aircraft in Flight: Special Report, 29 November 2002, para 4.38: http://www.rcep.org.uk/aviation/av12-txt.pdf

11. Office of National Statistics and Department for Transport, 'Car use in GB', Personal Travel Factsheet No. 7, January 2003, p. 4.

12. Department of Transport, standard constituent letter, 22 December 2005, personal communication.

13. Intergovernmental Panel on Climate Change, *Aviation and the Global Atmosphere: Executive Summary*, 2001: http://www.grida.no/climate/ipcc/aviation/064.htm

14. Royal Commission on Environmental Pollution, op. cit. (note 10, above), para 3.41.

15. For example, Malcolm V. Lowe, 'The Concorde May be History, but the Dream of Nonmilitary Supersonic Flight Still Lives', 16 November 2004: http://www.popularmechanics.com/science/aviation/1303021.html

16. Royal Commission on Environmental Pollution, op. cit. (note 10, above), para 5.28.

17. Department of Trade and Industry, *Energy: Its Impact on the Environment and Society*, 2005, p. 23: http://www.dti.gov.uk/files/file20263.pdf

18. Office of National Statistics, cited by Paul Brown and Larry Elliott, 'Air Travel Mars UK's Green Strategy', *Guardian*, 20 May 2005.

19. Office of National Statistics, 'Greenhouse Gas Emissions: Down since 1990', 19 May 2005:

http://www.statistics.gov.uk/CCI/nugget.asp?ID=901&Pos=6&ColRank=2&Rank=192

20. Department for Transport, White Paper: 'The Future of Air Transport', December 2003, p. 23:

http://www.dft.gov.uk/stellent/groups/dft_aviation/documents/page/dft_aviation_031516.pdf

21. ibid., p. 150.

22. ibid., p. 39.

23. ibid., p. 39.

24. ibid., p. 10.

25. ibid., p. 150.

26. ibid., p. 154.

27. Royal Commission on Environmental Pollution, op. cit. (note 10, above), para 2.24.

28. For example, Friends of the Earth, 'Aviation and the Economy', May 2005:
http://www.foe.co.uk/resource/briefings/aviation_and_the_economy.pdf

29. House of Commons Environmental Audit Committee, 'Environmental Audit – Third Report', 2004, para 12:

http://www.publications.parliament.uk/pa/cm200304/cmselect/cmenvaud/233/23305.htma5

30. ibid., para 13.

31. John Vidal, 'Heathrow Protesters Seek Help of Direct Action Group', *Guardian*, 20 February 2006.

32. Alice Bows, Paul Upham and Kevin Anderson, 'Growth Scenarios for EU and UK Aviation: Contradictions with Climate Policy', report for Friends of the Earth Trust Ltd, Tyndall Centre for Climate Change, 16 April 2005:

http://www.foe.co.uk/resource/reports/aviation_tyndall_research.pdf

33. ibid., pp. 10–11.

34. ibid., p. 12.

35. See Royal Commission on Environmental Pollution, op. cit. (note 10, above), paras 4.17–4.35.

36. Alice Bows and Kevin Anderson, 'Planned Growth in Aviation Contradicts Government Carbon Targets', *Climate Change Management*, Vol. 12 (January 2004).

37. Intergovernmental Panel on Climate Change, op. cit. (note 13, above).

38. Royal Commission on Environmental Pollution, op. cit. (note 10, above), para 3.41.

39. Department for Transport, op. cit. (note 20, above), p. 172.

40. House of Commons Environmental Audit Committee, op. cit. (note 29, above), para 15.

41. Civil Aviation Authority, 2004, cited by Transport 2000 in 'Facts and Figures: Aviation': http://www.transport2000.org.uk

42. MORI poll conducted for Freedom to Fly, January 2002, cited by Friends of the Earth *et al.*, in 'Aviation: The Plane Truth':
http://www.foe.co.uk/resource/factsheets/aviation_myths.pdf

43. HACAN/Clear Skies, 2005, cited by Transport 2000 in 'Facts and Figures: Aviation': http://www.transport2000.org.uk/

44. Department for Transport, op. cit. (note 20, above), p. 40.

45. House of Commons Environmental Audit Committee, op. cit. (note 29, above), para 49.

46. ACARE, cited by House of Commons Environmental Audit Committee, op. cit. (note 29, above).

47. *Decarbonising the UK – Energy for a Climate Conscious Future* (The Tyndall Centre for Climate Change Research, 2005), p. 50:
http://www.tyndall.ac.uk/media/news/tyndall_decarbonising_the_uk.pdf

48. ibid.

49. Royal Commission on Environmental Pollution, op. cit. (note 10, above), para 4.21.

50. ibid., paras 3.48 and 4.22.

51. Ulrich Schumann, 27 November 2000. 'Report on the European Workshop "Aviation, Aerosols, Contrails and Cirrus Clouds" (A^2C^3), Seeheim near Frankfurt/Main, July 10–12 2000', 27 November 2000:
http://www.aero-net.org/a2c3/a2c3_summary.pdf

52. Avions de Transport Régional, 'The Latest Generation Turboprop: The Green Power of Tomorrow', no date:
http://www.ataircraft.com/downl/The%20green%20power%20of%20
tomorrow.pdf

53. Royal Commission on Environmental Pollution, op. cit. (note 10, above), para 4.15.

54. Department for Transport, op. cit. (note 12, above).

55. Royal Commission on Environmental Pollution, op. cit. (note 10, above), para 4.16.

56. ibid., para 4.32.

57. op. cit. (note 47, above), p. 50.

58. Royal Commission on Environmental Pollution, op. cit. (note 10, above), para 3.43.

59. Bob Saynor, Ausilio Bauen and Matthew Leach, Centre for Energy Policy and Technology, 'The Potential for Renewable Energy Sources in Aviation', 7 August 2000, Imperial College, London:
http://www.iccept.ic.ac.uk/pdfs/ PRESAV%20final%20report%2003Sep03.pdf

60. ibid., p. 20.

61. ibid., p. 23.

62. Royal Commission on Environmental Pollution, op. cit. (note 10, above), 4.27.

63. ibid., para 3.47.

64. Intergovernmental Panel on Climate Change, *Aviation and the Global Atmosphere: Summary for Policymakers*, 2001:
http://www.grida.no/climate/ipcc/aviation/010.htm

65. Department for Transport, op. cit. (note 20, above), p. 40.

66. *Railway Gazette International*, '2005 World Speed Survey Tables', 1 December 2005:
http://www.railwaygazette.com/Articles/2005/11/01/1222/2005+World+Speed+Survey+tables.html

67. Central Japan Railway Company, 'Successful Development of the World's Highest-performance High-temperature Superconducting Coil', no date:
http://jr-central.co.jp/eng.nsf/english/bulletin/$FILE/vol46-tokai.pdf

68. *Railway Gazette International*, op. cit. (note 66, above).

69. For example, Martin Wainwright, 'Hovertrain to Cut London–Glasgow Time to Two Hours', *Guardian*, 9 August 2005.

70. UK Ultraspeed, '500km/h Ground Transport for Britain', no date:
http://www.500kmh.com/index.html

71. Central Japan Railway Company, 'Chuo Shinkansen', no date:
http://jr-central.co.jp/eng.nsf/english/chuo_shinkansen

72. anon., 'Shanghai Maglev Gets Official Nod of Approval', *China Daily*, 27 April 2006: http://www.china.org.cn/english/China/166911.htm

73. Eddie Barnes, 'Scottish Deputy First Minister Backs Glasgow–Edinburgh Link', *Scotland on Sunday*, 27 November 2005.

74. http://en.wikipedia.org/wiki/Magnetic_levitation_train

75. UK Commission for Integrated Transport, 'High-speedRail: International Comparisons', 9 February 2004, Figure 4.1, p. 33:
http://www.cfit.gov.uk/docs/2004/hsr/research/pdf/chapter4.pdf

76. Roger Kemp, 'Transport Energy Consumption', Lancaster University, 10 September 2004:
http://www.engineering.lancs.ac.uk/research/download/Transport%20Energy%20Consumption%20Discussion%20Paper.pdf

77. Roger Kemp, 'Environmental Impact of High-speed Rail', Institution of Mechanical Engineers, 21 April 2004:
http://www.engineering.lancs.ac.uk/research/download/Environmental%20impact.pdf

78. Martin Wainwright, op. cit. (note 69, above).

79. Roger Kemp, op. cit. (note 77, above).

80. http://www.bgtgroup.com/marine_turbine.htm

81. http://www.cunard.com/OnBoard/default.asp?OB=QE2&sub=sp

82. George Marshall, personal communication.

83. 'Four 8,000 kW diesel engines propelling fewer than 1000 passengers [gives] (at best) 4 x 8000 / (1000x30) = 1kWh/passenger mile. A similar calculation for the London–Edinburgh trains which have around 500 seats and use 3,000kW at 125mph gives 3000 / (500x125) = 0.05kWh/passenger-mile.' Roger Kemp, personal communication.

84. Tim Thwaites, 'Slippery Ships Float On Thin Air', *New Scientist*, 18 February 2006.

85. http://skysails.info/index.php?L=1

86. Alice Bows, Kevin Anderson and Paul Upham, *Contraction and Convergence:*

UK Carbon Emissions and the Implications for UK Air Traffic, Tyndall Centre Technical Report No. 40, February 2006, p. 23:
http://www.tyndall.ac.uk/research/ theme2/final_reports/t3_23.pdf

87. Kevin Anderson, personal communication.

88. http://www.hacan.org.uk

89. http://www.planestupid.com

10 VIRTUAL SHOPPING

1. *Faust*, page 42.

2. George Monbiot, *Captive State: The Corporate Takeover of Britain* (Macmillan, London, 2000), p. 176.

3. Nigella Lawson, *How to Eat* (Chatto and Windus, London, 1998), p. 47.

4. Robert Hogg and Henry Graves Bull, *The Herefordshire Pomona* (Jakeman & Carver, Hereford, 1876).

5. *Henry IV*, Part 1, Act III, Scene III.

6. Royal Commission on Environmental Pollution, 'Biomass as a Renewable Energy Source', 2004, Table 3.1, p. 38:
http://www.rcep.org.uk/biomass/Biomass%20Report.pdf

7. J Sainsbury PLC, 'Sainsbury's Breathes Life into Greenwich Peninsula', press release, 14 September 1999:
http://www.j-sainsbury.co.uk/files/reports/er1998/whatsnew.htm

8. ibid.

9. Calls and e-mails from Matthew Prescott to J Sainsbury, 20 December 2005, 29 December 2005 and 9 January 2006.

10. J Sainsbury PLC, op. cit. (note 7, above).

11. ibid.

12. Nick Martin, 'Can We Harvest Useful Wind Energy from the Roofs of Our Buildings?', *Building for a Future*, winter 2005/6, special wind-power feature, Table 2.

13. Department for Transport, 'Transport Statistics Bulletin', 2005, Section 7, Table 7.1:
http://www.dft.gov.uk/stellent/groups/dft_transstats/documents/page/dft_trans stats_039338.pdf

14. ibid., Table 7.2.

15. Greg Palast, *The Best Democracy Money Can Buy* (Constable and Robinson Ltd, London, 2003), pp. 279–80, 285.

16. Department of the Environment, Transport and the Regions, *A New Deal for Transport: Better for Everyone* (The Stationery Office, London, 1998).

17. S. Cairns *et al.*, *Smarter Choices – Changing the Way We Travel*, Department for Transport, 2004, Chapter 12 (Home Shopping):
http://www.dft.gov.uk/stellent/groups/dft_susttravel/documents/page/dft_susttravel _029756.pdf

18. ibid., p. 305.

19. Tesco Customer Services (08457 225533), 6 April 2006, personal communication.

20. http://www.tesco.com/help/page.asp?choiceA=groc&choiceB=5

21. Rob Scott McLeod, 'Ordinary Portland Cement', *Building for a Future*, autumn 2005.

22. Ernst Worrell *et al.*, 'Carbon Dioxide Emissions from the Global Cement Industry', *Annual Review of Energy and Environment*. Vol. 26 (2001), pp. 303–29.

23. ibid.

24. David Ireland, 'The Green House Effect', *Guardian*, 5 May 2005.

25. Ernst Worrell *et al.*, op. cit. (note 22, above).

26. Rob Scott McLeod, op. cit. (note 21, above).

27. Fred Pearce, 'The Concrete Jungle Overheats', *New Scientist*, 19 July 1997.

28. David Pocklington, British Cement Association, personal communication.

29. Fred Pearce, op. cit. (note 27, above).

30. Ernst Worrell *et al.*, op. cit. (note 22, above).

31. Jon Gibbins *et al.*, 'Scope for Future CO_2 Emission Reductions from Electricity Generation through the Deployment of Carbon Capture and Storage Technologies', paper presented to the 'Avoiding Dangerous Climate Change' conference, Meteorological Office, Exeter, 1–3 February 2005.

32. Intergovernmental Panel on Climate Change, *Climate Change 2001: Working Group III –Mitigation*, Section 3.5.3.5: Material Efficiency Improvement: http://www.grida.no/climate/ipcc_tar/wg3/109.htm

33. Rob Scott McLeod, op. cit. (note 21, above).

34. Autoclaved Aerated Concrete, 'The Material Aircrete (Autoclaved Aerated Concrete, AAC) and How to Make It, no date: http://www.pb-aac.de/basemat.html

35. David Olivier, Energy Advisory Associates, letter to *Building for a Future*, winter 2005/6.

36. Godfrey Boyle, Bob Everett and Janet Ramage (eds.), *Energy Systems and Sustainability* (Oxford University Press, Oxford, 2003), Box 3.2, p. 106.

37. Portland Cement Association, 'High Strength Concrete', no date: http://www.cement.org/basics/concreteproducts_histrength.asp

38. D. J. Gielen, 'Building Materials and CO_2: Western European Emission Reduction Strategies', Netherlands Energy Research Foundation, 1997, p. 55: http://www.ecn.nl/docs/library/report/1997/c97065.pdf

39. ibid.

40. ibid., p. 60.

41. Michael Judge, 'Our Flexible Friend', *New Scientist*, 10 May 1997.

42. Gayle M. B. Hanson, 'Like a Rock . . . Only Harder: Process for Making New Building Material', *Insight on the News*, 4 August 1997.

43. Michael Judge, op. cit. (note 41, above).

44. Supramics, 'Better, Cheaper, Faster . . . and Cleaner', press release, 21 May 1996: http://www.supramics.com/press/supress1.html

45. Michael Judge, op. cit. (note 41, above).

46. Supramics, op. cit. (note 44, above).

47. Phone conversation with Roger Jones, Supramics, 7 April 2006.

48. Los Alamos National Laboratory, 'Los Alamos Paves the Way for Better Cement', 27 January 2007: http://www.supramics.com/press/lanl.html

49. TecEco Pty Ltd, 'Eco-cement', no date: http://www.tececo.com/simple.eco-cement.php

50. John Harrison, TecEco, quoted by Fred Pearce, 'Green Foundations', *New Scientist*, 13 July 2002.

51. Benjamin Herring, 'The Secrets of Roman Concrete', *Constructor*, September 2002: http://www.romanconcrete.com/Article1Secrets.pdf

52. Geopolymer Institute, 'Cements, Concretes, Toxic Wastes, Global Warming', no date: http://www.geopolymer.org/what_is_a_geopolymer/geopoly_vs_portland_cement.html

53. Zongjin Li, Zhu Ding and Yunsheng Zhang, 'Development of Sustainable Cementitious Materials', in *Proceedings of the International Workshop on Sustainable Development and Concrete Technology*, Beijing, China, May 2004, pp. 55–76: http://www.cptechcenter.org/publications/sustainable/lisustainable.pdf

54. Rob Scott McLeod, op. cit. (note 21, above).

55. Geopolymer Institute, op. cit. (note 52, above).

56. CSIRO, 'Geopolymers: Building Blocks of the Future', no date: http://www.csiro.au/csiro/content/standard/ps19e,,.html

57. Zongjin Li, Zhu Ding and Yunsheng Zhang, op. cit. (note 53, above).

11 APOCALYPSE POSTPONED

1. *Faust*, page 33.

2. David R. Criswell, University of Houston, 'Lunar Solar Power System: Industrial Research, Development and Demonstration', World Energy Council 18th Congress, Buenos Aires, October 2001: http://www.worldenergy.org/wec-geis/publications/default/tech_papers/18th_Congress/dsessions/ds2/ds2_17.asp

3. For example, A. Gnanadesikan, J. L. Sarmiento and R. D. Slater, 'Effects of Patchy Ocean Fertilization on Atmospheric Carbon Dioxide and Biological Production', *Global Biogeochemistry Cycles*, Vol. 17 (2003), p. 1050.

4. The Royal Commission on Environmental Pollution, *Energy – The Changing Climate*, June 2000, para 3.26: http://www.rcep.org.uk/newenergy.htm

5. Marina Murphy, 'The Wright Wind Scrubbers: First Prototype', *Chemistry and Industry*, 5 April 2004: http://www.chemind.org/CI/viewissue.jsp?dateOfIssue=2004–04-05&currentIssueType=News&pt=CI

6. Michael Behar, 'How Earth-scale Engineering Can Save the Planet', *Popular Science*, June 2005: http://www.popsci.com/popsci/aviationspace/3afd8ca927d05010vgnvcm1000004eecbccdrcrd/4.html

7. Paul Rincon, 'Plan to Build emissions scrubber', *BBC News Online*, 13 April 2004: http://news.bbc.co.uk/1/hi/sci/tech/3612739.stm

8. Liz Tilley, Global Research Technologies, 11 April 2006, personal communication.

9. anon., 'A Mirror to Cool the World', *New Scientist*, 27 March 2004.

10. Michael Behar, op. cit. (note 6, above).

11. anon., op. cit. (note 9, above).

12. Kenneth Deffeyes, quoted by Bob Holmes and Nicola Jones, 'Brace Yourself for the End of Cheap Oil', *New Scientist*, 2 August 2003.

13. US Geological Survey, 'World Petroleum Assessment 2000: Executive Summary', 2000: http://pubs.usgs.gov/dds/dds-060

14. Matthew Simmons, *Twilight in the Desert: The Coming Saudi Oil Shock and the World Economy* (Wiley, New York, 2005).

15. Energy Intelligence, 'High Oil Prices: Causes and Consequences', 2005: http://www.energyintel.com/datahomepage.asp?publication_id=65

16. Jean Laherrere, 'Is USGS 2000 Assessment Reliable?', 2 May 2000: http://energyresource2000.com

17. Robert L. Hirsch, Roger Bezdek and Robert Wendling, 'Peaking Of World Oil Production: Impacts, Mitigation, and Risk Management', US Department of Energy, February 2005, available at http://www.hubbertpeak.com/us/NETL/OilPeaking.pdf

18. John Lothrop Motley, *The Rise of the Dutch Republic*, 1855, Part 2, Chapter 11, available at http://historicaltextarchive.com/books.php?op=viewbook&bookid=60&cid=11

19. Larry Lohmann, 'Marketing and Making Carbon Dumps: Commodification, Calculation and Counterfactuals in Climate Change Mitigation', *Science as Culture*, Vol. 14 (September 2005), pp. 203–35.

20. See Kevin Smith *et al.*, *Hoodwinked in the Hothouse: The G8, Climate Change and Free Market Environmentalism*, Transnational Institute Briefing Series, 30 June 2005: http://www.tni.org/reports/ctw/hothouse.pdf

21. Esteve Corbera, 'Bringing Development into Carbon Forestry Markets: Challenges and Outcomes of Small-scale Carbon Forestry Activities in Mexico', in D. Murdiyarso and H. Herawati (eds.), *Carbon Forestry: Who Will Benefit?* (Center for International Forestry Research, Bogor, Indonesia, 2005).

22. Robert B. Jackson *et al.*, 23rd December 2005. 'Trading Water for Carbon with Biological Carbon Sequestration', *Science*, Vol. 310 (23 December 2005), pp. 1944–7.

23. Mark Broadmeadow and Robert Matthews, 'Forests, Carbon and Climate Change: The UK Contribution', June 2003, p. 3: http://www.forestry.gov.uk/pdf/fcin048.pdf/$FILE/fcin048.pdf

Acknowledgements

I could not have written this book without the help of the following people. My wife Angharad Penrhyn Jones, my researcher Dr Matthew Prescott, my assistant Sandy Kennedy, my agents Antony Harwood and James Macdonald Lockhart, and my editor Helen Conford. I've also relied heavily on the following experts, who have been very generous with their time, in this order: Dave Andrews, Graham Sinden, Robert Gross, Simon Bowens, George Marshall, Mark Lynas, Andrew Warren, Oliver Tickell, Richard Starkey, Stephen Joseph, Paul Allen, Roger Kemp, Brenda Boardman, Becca Lush, Byron Wine, Neil Crumpton, Norbert Hirschhorn, Kevin Anderson, Norman Myers, Stuart Croft, Sammy Daniel, Sarah Keay-Bright, Alice Bows, Jonathan Gibbins, Rajat Gupta, David Hirst, Walt Patterson, Meyer Hillman, Andrew Simms, Thomas Aldrian, Mark Barrett, Robert Webb, Dieter Helm, Lex Waspe, Jim Skea, John Loughhead, Les Harding, Mark Partington and Ian Gazley. Many thanks indeed to all of you.

Index